John Alexander
3, Greenbanks Drive

Barry

Tele 736364.

1972.

978 0851130002

D1806325

More Mini tuning

by Clive Trickey

SPEED SPORT
MOTOBOOK

First impression November 1968
Second edition October 1970

(85113-000-3)

Printed and published by
Speed and Sports Publications Ltd.,
Acorn House, Victoria Road, Acton, London W3.

© Copyright 1970—Speed and Sports Publications Ltd.
All rights reserved.

Contents

chapter one

Carburetters

The number and variety of carburetters offered on the market as tuning aids to the Mini are legion. Some are more successful than others. Let us take a look at a specific SU carb, namely the $1\frac{1}{2}$in H4 SU and also the $1\frac{1}{2}$in Reece-Fish carb. The reason for dealing with these two particular carburetters is the very popular formula Mini 7 class of saloon car racing. More and more people seem to be preparing cars to these regulations and since a single carb with a maximum choke tube diameter of $1\frac{1}{2}$in is mandatory, many people have had carburation problems. The H4 and the Reece-Fish have emerged as the two most popular instruments to use, the former due to its cheapness and ready availability second hand, coupled to simplicity and the excellent results that can be obtained when properly set up. The Reece-Fish has emerged as at present being 'The complete answer' to carburation for the formula, due of course to the excellent bonus in power and performance which it offers. It is also very simple and in many respects even easier to tune than an SU.

The H4 SU carburetter
Why do I suggest using the H4 SU instead of the $1\frac{1}{2}$in HS4? Well, the flange which fixes to the manifold is more universal; it can be fixed to the same stud fixings as a standard Mini HS2 carb. Should you wish to use a standard inlet manifold for economy reasons, second-hand manifolds to suit this carb are more plentiful and most important in my opinion, if it is the intention to change to Reece-Fish later you need only change carburetters, the manifold fixings are

7

identical. If you use an HS4 carb you will need to change manifolds, all adding to the expense.

As regards performance, the H4 has a slightly longer choke tube than the HS4 and it is my opinion that this aids the gas-flow through the carburetter by increasing the gas velocity. This is only marginal, but even margins are important when racing.

Before trying to sort the mixture, you must first get the carb set-up properly, carrying out all required modifications etc. If you assemble an SU carb and raise the piston as far as you can with your finger, you will usually find that even at full lift, as far away from the bridge as it will go, the piston still partially projects into the choke tube, thus effectively reducing its diameter. The reason is that quite often the damper tube, which is part of the piston, is too long, and at full lift fouls against the screwed damper cap on the top of the piston cover preventing the piston rising any further. The remedy is brutal but simple: shorten the tube on the piston by sawing or grinding an appropriate amount off the top. Remember, however, that you must remove any resultant burrs from inside or outside the tube, as these will otherwise prevent the piston sliding up and down freely and this would be disastrous to performance.

Next, modify the piston itself. Firstly it should be of the quick-lift variety, obtainable direct from SU or possibly from your local speed shop or engine tuners. Unfortunately many people seem to have difficulty in obtaining these and even my old source has dried up. They say that it is no longer profitable, though no doubt if you went along to your local engine tuners armed with your standard piston and a couple of pound notes they would do the job for you. If you have trouble try Janspeed Engineering at Salisbury.

It is possible to do the job yourself. The existing hole in the side of the piston should be blanked off and two new holes $\frac{1}{8}$in diam and slots $\frac{3}{32}$in deep put into the underside of the piston as shown in the diagram.

The lower leading edge of the piston, that is the edge away from the

manifold towards the air intake should be nicely radiused off for about $\frac{3}{8}$ of its circumference, the greater radius being in line with the longitudinal centre line of the choke tube tapering off to nothing either side, stopping just behind the bridge. No attempt should be made to radius the manifold side of the piston, or the bridge itself. Make the greatest part of the radius a radius of about $\frac{3}{8}$in.
The mouth or air intake of the carb can also be radiused on the same radius for the whole of its circumference. Ordinary hand tools can be used for this, a final polish being obtained with wet/dry paper.

The butterfly can have its leading edge—towards air intake— knife-edged and the spindle diameter can be almost halved. Butterfly fixing screws should be cut-off flush with the spindle.

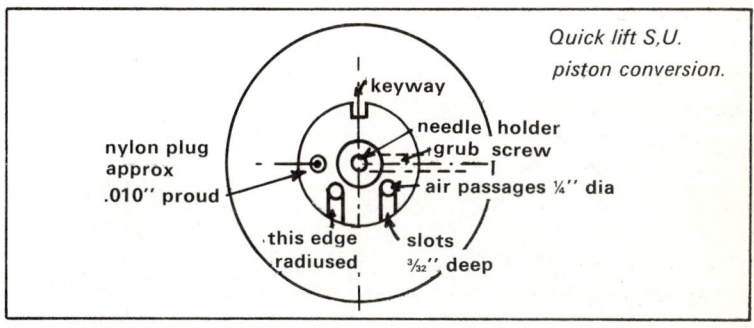

Quick lift S.U. piston conversion.

keyway
needle holder
grub screw
air passages ¼″ dia
slots ³⁄₃₂″ deep
this edge radiused
nylon plug approx .010″ proud

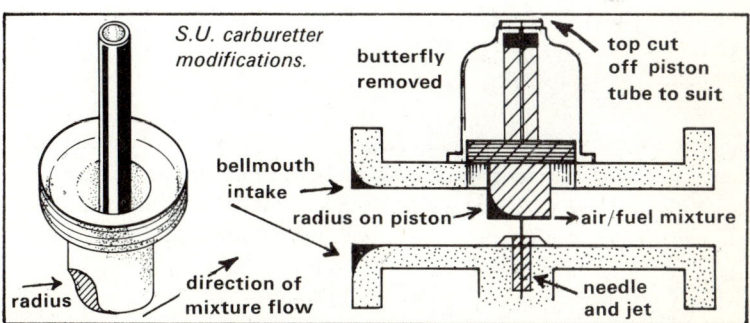

S.U. carburetter modifications.
butterfly removed
top cut off piston tube to suit
bellmouth intake
radius on piston
air/fuel mixture
needle and jet
radius
direction of mixture flow

The linkage can be modified so that the butterfly opens 'the other way round'. That is instead of the bottom half of the butterfly moving towards the air intake and the top half towards the manifold, you make it vice-versa. This will give slightly better throttle response.

Most people find that a full-race 850 Mini fitted with 649 camshaft and large valve head requires a very rich mixture. Even SU's .090 needle, the BG, is too weak. All you can do is make the needle slightly thinner over the bottom $\frac{1}{4}$ of its length (i.e. where it is already thinnest) to richen the mixture for higher engine rpm where it is needed. I can't tell you how much; the full-race engine is very susceptible to even small changes in engine or exhaust system specification. All you can do is experiment and arrive at the right answer by trial and error, using plug tests as a guide. You will probably find that if you stick to recommended float chamber fuel levels, the car will be almost undrivable on right hand corners, at least at racing speeds. The engine will cut dead. Most unpleasant, indeed, very dangerous. There are three ways of overcoming this.

I found that the simple way was quite effective. The float control arm was bent up so that it did in fact touch the float chamber lid, and the needle valve never closed. Consequently, if the ignition was switched on but the engine not started, the whole lot flooded. If the car was left parked facing downhill, petrol siphoned straight out of the tank through the float bowl breather tube on to the ground. Not exactly ideal for a road car! But this was an out-and-out racer, the side effects were tolerable, they disappeared when the engine was running and I overcame my fuel surge problems.

The second answer is to fit an additional float bowl, so that you have one either side of the main carburetter body. The additional bowl is simple enough to fit, using a long double-drilled fixing bolt with two fuel feed holes. Unfortunately you will find the throttle linkage a real nuisance to arrange. Being lazy, I did not use this method, even though I had the necessary bits.

Recently the SU carb manufacturers have themselves been marketing a special extended float chamber lid which increases the overall height of the float chamber and enables the float to be set at a much higher level. If you can obtain such a chamber lid it probably represents the best remedy to the age-old surge problem, certainly it is very easy to fit being a straight substitute for the standard unit.

Well there it is, a simple effective cheap carb. However I now use a Reece-Fish and reckon that it is much better from a power standpoint. However, don't let this put you off racing if you can't afford a Reece-Fish. You can still go very quickly on an SU. You can have a lot of fun and the knowledge that you still have a fair bit of development left is always most comforting.

The Reece-Fish carburetter
The 1½in Reece-Fish carburetter is at present regarded by most people, myself included, as the ultimate in carburation on a formula Mini 7 racing saloon.

Needless to say, the fact that I am considering these carburetters in the light of Mini racing experiences does not mean that this is their sole application; they are most effective on almost any road or rally production car, whether or not the engine is otherwise modified. There is, however, one big difference between the H4 SU and the Reece-Fish as regards application: the SU only really gives better

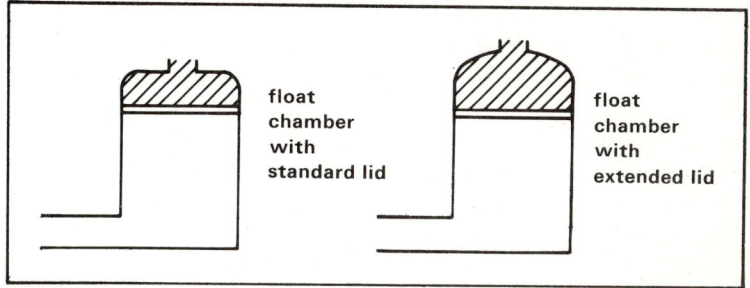

float chamber with standard lid

float chamber with extended lid

results—when compared with a standard fitting—on the BMC A series engine, and then only on engines up to 1,100cc. Thus its use—as a single instrument—as a tuning *extra* is somewhat limited. In my opinion no one in their right mind would think of fitting a single H4 SU in place of the standard Weber carburetter on a Cortina GT. On the other hand the 1½in Reece-Fish gives excellent and improved results on almost any engine up to 2 litres, including the Cortina GT, where it is used to replace the more mundane single-choke or progressive twin-choke commercial setups. In other words the Reece-Fish not only seems better than the SU on those engines on which a 1½in SU works well, but it can also be used with advantage on much larger engines, even where each cylinder has a separate inlet port. It is far more universal and, used singly, often gives better results than twin carburetters or multiple choke set-ups although this latter phenomenon is more pronounced on small engines up to 1,300cc, particularly where siamesed inlet ports are used. The Reece-Fish comes in two forms, down-draught and semi-down-draught. From a tuning viewpoint they can be regarded as identical, the former being used where it replaces a down-draught carburetter such as Solex, Zenith, Fomoco, 28/36 DCD Weber, etc. The semi-downdraught is used to replace such carbs as the SU or Zenith-Stromberg on inclined manifolds. Where a horizontal manifold is employed one needs to use an adaptor plate, when one can use either the semi- or full downdraught versions. At present these carbs come in two readily available sizes, 1¼in and 1½in. For the moment it's the 1½in that I am interested in, though the principles and description of operations could equally apply to any of the sizes in the range. In describing the mode of operation and tuning procedure I can do no better than quote word for word the makers' own hand-out as follows:

"Apart from having a float feed, the Reece-Fish carburetter is entirely different from other types as it has no choke tube, taper needle, jets or air strangler. The fuel level is relatively unimportant and the only moving parts are the throttle spindle, needle valve and leaf valve. The float chamber is comprised of two compartments separated by a metal diaphragm which carries a leaf valve—one containing the float, whilst the other houses the radial fuel pick-up arm attached to the throttle spindle. The fuel is metered for the relative throttle openings by a horizontal hole in the pick-up arm, which registers with a calibrated groove machined radially in the chamber wall, through the pick-up arm, and is discharged through the orifices in the throttle spindle. As the pick-up arm fits closely into the inner chamber, it also acts as a displacer or piston. When the throttle is opened sharply, the leaf valve closes and the chamber becomes pressurised, forcing fuel through the arm and spindle into the air stream for an acceleration shot.

The carburetter should be fitted with the float chamber facing the front of the car. On an inline engine, where an inclined or horizontal carburetter is being replaced, a left- or right-hand model must be used. This does not apply with a transverse engine. In order to eliminate the risk of heavy loading and subsequent wear on the main

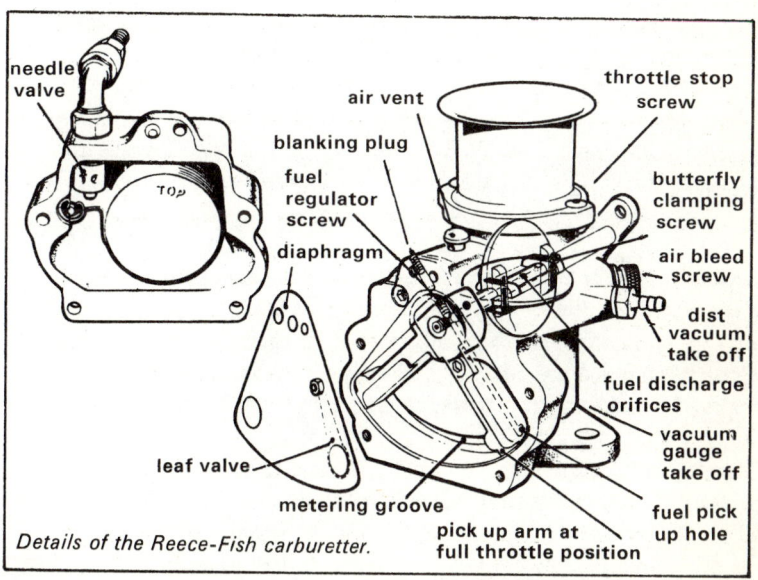

Details of the Reece-Fish carburetter.

spindle, the throttle control linkage should operate direct and without bias, preferably through a bell crank or telescopic return spring and no opposing return springs should be used on the throttle arm.

Each carburetter is bench tested and, where possible, adjusted to the engine capacity for which it is intended. However, individual tuning is still necessary after fitment and can be divided into two stages, maximum power and cruising range.

Maximum power

Maximum power is obtained by using the socket key provided and adjusting the fuel flow regulator screw in the fuel pick-up arm with engine running at full throttle under load. Turn clockwise to weaken and the reverse to enrich until the optimum is reached. As can be seen in the diagram, this screw is only accessible at the full throttle position after the blanking plug has been removed.

This operation should be carried out on a roiler dynamometer for the quickest and best results. Alternatively, adjustment can be made on the road by using an uphill section of about half a mile, approaching a roadside marker at 40mph and opening the throttle fully as the marker is passed, noting the speedometer reading at two or more points on the climb. With the engine stopped, the mixture strength should now be weakened by turning the regulator screw one "flat" clockwise on the hexagonal key. Replace blanking screw.

Repeat run under exactly the same conditions, again noting the speedometer readings. If an improvement is shown, weaken the mixture still further by the same amount. If there is no improvement, it can be assumed the mixture is already too weak and must be enriched by turning the regulator screw two "flats" anti-clockwise, i.e. one "flat" beyond the initial setting.

Continue runs, adjusting one "flat" at a time until power is noticed to drop off, then return regulator screw to previous best setting. These runs may be taken on a level road, remembering to hold the throttle fully open for the length of the run after passing the first marker. Having attained maximum power, weaken the mixture strength slightly, less than one "flat" to ensure maximum economy together with peak performance.

Cruising range

For maximum output and peak efficiency of an engine, the ideal air/fuel ratio of 15 to 1 should remain constant throughout the throttle range. As peak efficiency produces the highest manifold vacuum reading, it is obvious, once such reading is obtained, the air/fuel ratio is correct. Therefore, by altering the relationship of the butterfly controlling the air and the pick-up arm controlling the quantity of fuel metered by the groove, radially upon the spindle until the highest vacuum reading is obtained, the correct air/fuel ratio and peak efficiency will be assured. A good quality vacuum gauge must be used when making the adjustment:

Close air bleed screw against seating and open half turn.

Remove plug and connect vacuum gauge.

Start engine and run to normal working temperature.

Partly loosen butterfly clamping screw to allow restricted radial movement using key.

Adjust throttle stop screw for engine to run 2000 rpm.

Using scriber, block centre hole on butterfly spindle—note effect on vacuum gauge reading.

Vacuum rise indicates weak mixture and can be corrected by closing the butterfly slightly, thereby admitting less air for the same amount of fuel. This can best be effected with a light tap on the upper half of the butterfly using the blunt end of the scriber.

Vacuum fall indicates a rich mixture. In this case it is necessary to admit more air by opening the butterfly—tapping the lower half.

Reset engine speed to 2000 rpm.

Repeat operation until vacuum gauge reading remains constant as the centre hole in the spindle is blocked.

Tighten butterfly clamping screw.

Check gauge reading using scriber as before to ensure butterfly did not move when tightening screw.

Adjust throttle stop screw for tick-over and, if necessary, alter air bleed screw.

Cruising range adjustment should always be checked and reset if any alteration is made to the maximum power mixture adjustment.

Ignition timing can generally be advanced some five or more degrees beyond standard setting. Adjustments should be made progressively and tested under similar conditions as used for maximum power.

Cold starting

No hard and fast rule can be given, but it is usually best effected by sharply depressing the accelerator pedal once or twice to prime the manifold, closing the throttle completely and operating the starter. The engine should start immediately and may be assisted by gentle pumping on the accelerator until even running and a fast tickover is obtained, which should be held a few moments before moving off.

When making any adjustments, the throttle should always be opened very slowly to reduce the efficiency of the fuel pump, otherwise the manifold will become flooded and subsequent starting made difficult. Opening the throttle fully and operating the starter will usually rectify this.

Cleanliness of the fuel supply is of utmost importance—an additional filter should be fitted wherever possible. By removing the float chamber and the diaphragm, the "innards" can be cleaned by the use of an airline. No attempt should be made to dismantle the carburetter further as no useful purpose will be served. Care must be taken when replacing the diaphragm that it settles into the register provided. To ensure freedom from obstruction, fuel can be pumped through the spindle into a container after the carburetter has been removed from the engine. On transverse engines, a mirror held in the left hand, and used as by a dentist, is helpful in locating the butterfly clamping screw and centre hole on the spindle.

On a racing car, road testing is very difficult and although one can substitute track testing I found that by far the simplest and quickest method was to scrounge the use of the local Austin distributor's excellent 'Rolling-Road', manufactured by Crypton. A final check on mixture can be made by the usual plug check.

For over a year I had no reason to doubt the validity of such a tuning procedure, after all these carbs were in everyday use by hundreds of people, and dozens were successfully used on 850 Mini 7 racers. We had all followed this same tuning procedure and found it completely satisfactory. Certainly the carburation on my KTR 223E during 1967 had been perfect, the car going like stink as a result.

However, came 1968 and KTR had a different engine with a somewhat altered cylinder head, necessitating a different mixture setting. Religiously I followed the hitherto accepted tuning procedure. At first, all seemed well, I had set the butterfly to give maximum vacuum and was progressively-richening the main mixture control grub screw. Lack of time to hire a circuit for long enough and awkward neighbours preventing the use of the rolling-road, meant that all experimenting had to be carried out at actual race meetings during the very short practice sessions. Nonetheless, each new practice session saw me with a slightly richer mixture setting—based on the grub screw setting—and a little extra power. Then came the day at Lydden September Bank Holiday meeting. This tuning procedure fell flat on its face. I had been richening the mixture because it was still far, far too weak at maximum rpm and was seriously restricting top-end power. Suddenly the mixture was too rich and as I put my foot down, the engine went on three cylinders, till revs rose to about 5000, when it came back on to four. But it wasn't too rich at peak revs, it was much too weak. Both exhaust pipe and plugs were white, and power dropped off badly at 7000rpm even in second gear. Don't ask me to explain how you can feel the difference between a misfire due to over richness, and a miss due to weakness. All I can say is that the former seems to make the engine growl, whilst the latter misses with a series of small backfire like reports.

You will know the difference once you have experienced these phenomena. Needless to say, I was at first very puzzled. I had followed the accepted sequence, but had reached an over-richness state without even approaching the correct mixture at peak rpm. Rather like using too small a carburetter though, I was convinced a 1½in Reece-Fish was plenty big enough. It seemed that the basic overall setting of the carb was too weak.

The butterfly position seemed to dictate the carb's *overall* setting and determine its general character. The mixture control grub screw merely metered fuel, the butterfly determined the air/fuel ratio. How could a setting which was perfect at 2000 or even 3000rpm be correct at 7500 to 8000? At very high rpm it appeared to me that gas velocity built up to a point where, coupled to the relative spindle drillings/butterfly dispositions, the fuel to air mixture ratio favoured the air and hindered the petrol, allowing the mixture to become too weak. I figured that though the traditional method was the answer for road engines or those not being taken above 7000rpm, it was no good for high revving racers. Even though the mixture control grub screw was metering enough fuel (in fact too much) into the spindle and thence out through the spindle drillings, the fuel was not in fact escaping through the drillings in sufficient quantities at maximum rpm due to the partial masking effect of the butterfly and the relative pattern of the air flow over it.

The answer seemed to me to richen the butterfly setting. Obviously, however, extreme care was necessary in this. It was no use relying on a vacuum gauge to indicate the setting. But I still needed to know what setting the butterfly was at, relative to its earlier position. Only thus could I work out an orderly progression and sequence of operations. Then, Ken, who looks after the carburetters at the Reece establishment, had a great idea, as follows. If one loosens the butterfly on the spindle and rotates it free of the spindle, moving the butterfly into a closed position *richens* the mixture and opening it up *weakens* it. This gave us our method of controlling the degree to which the butterfly was altered from its previous position, without recourse to a vacuum gauge. In fact a far superior method, but there is a snag. Unlike the original theory of tuning, one cannot just slap the carb on and adjust its basic setting on a vacuum. It must be done by trial and error, but more of this in closing.

It is advisable to use a vacuum gauge to obtain a starting point, setting the butterfly for maximum depression. For low revving engines, this is all that is necessary. But for a racer, it will almost certainly prove too weak. However, you are now at a known point, which you can return to if need be. Next, start the engine and, without driving it, let it rev to about 7000rpm or better still, take it out and drive it. If it misses, pops, bangs, backfires and just won't go, then you know that the grub screw is set too weak. Screw it out, till you have overcome the misfire exactly as described in the manufacturer's hand-outs. Now, it's up to you to follow the new sequence as set below.

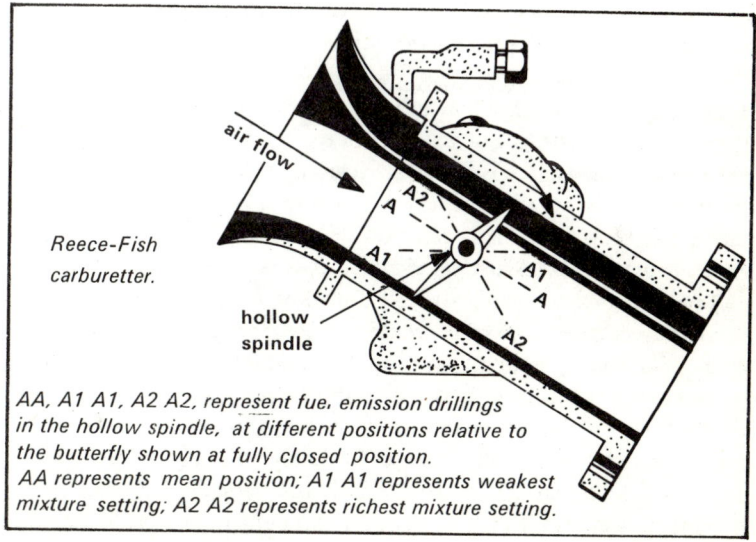

Reece-Fish carburetter.

AA, A1 A1, A2 A2, represent fuel emission drillings in the hollow spindle, at different positions relative to the butterfly shown at fully closed position.
AA represents mean position; A1 A1 represents weakest mixture setting; A2 A2 represents richest mixture setting.

(1) Re-adjust the throttle-stop control screw by turning it clockwise half a turn, which opens throttle slightly.

(2) Slacken butterfly clamping screw and press butterfly to fully closed position. Re-tighten clamp. Thus you have now enriched the mixture by an amount equivalent to $\frac{1}{8}$ turn on the throttle control screw.

This is all the new procedure involves. Unfortunately there is no simple and straightforward sequence whereby you can set A for optimum results and then concentrate on B separately till this is also set properly. With my new method, the butterfly and grub screw must be progressively altered together, till optimum overall results are obtained. If one reaches the point of over-richness on the grub screw, then weaken its setting, but continue enrichening on the butterfly. If the butterfly becomes too rich first, then weaken this setting and enrichen the grub screw. Keep going till alterations to either only reduce performance. It's a bit of a fiddle, very much trial and error, but anything worthwhile is worth extra effort. Obviously a dynamometer will save considerable time.

I have explained roughly how to recognize richness or weakness of the grub screw setting, but this is even easier on the butterfly. If set too rich and the engine is started and revved up, then when the throttle is closed, the engine will continue to rev for several seconds more, only slowly returning to an idle. Conversely, if too weak, the engine won't easily start. Over richness does not seem to affect top end performance too badly, at least providing it's not overdone. In fact racing engines need what would normally be regarded as an over-rich butterfly setting, almost to the point where run-on occurs when closing the throttle as just described. One point worth remembering is that an engine set-up such as this becomes a little more temperamental, difficult to set for idling and easy to juice plugs on when fiddling around in the paddock, but perfectly OK once moving on a circuit. Small sacrifice for extra engine power. Somehow last year I had accidentally set the butterfly too rich and by sheer fluke obtained the right setting, only needing to adjust the grub screw to obtain optimum overall performance.

One problem encountered on my racing Mini was common to the SU and the Reece-Fish, fuel surge. Fortunately this was only encountered at a practice day at Castle Combe and Mr Reece immediately came up with the answer. A special float chamber which, I understand will soon become a standard fitment on all his carburetters. It is wedge-shaped, being wider at the bottom than the top. When fuel surges it moves away from the base of the chamber away from the fuel pick-up and into the roof of the chamber, literaly climbing up the chamber wall. By enlarging the bottom of the chamber it can be made to hold more fuel overall, but more important, since the extra volume is at the bottom, it obviously holds a greater volume of fuel than the top part of the chamber. Thus when surge takes place the narrower top half of the chamber is soon filled and no more fuel can surge from bottom to top, thus preventing the pick-up becoming starved, and eliminating any unpleasant effects. I still found however that I needed to raise the fuel level, by bending the float control arm, to a point where the carburetter just flooded if the ignition was left switched on, the engine not running, for about 45 seconds. This was all that was necessary, but on the early, more rectangular type, one needs to raise the fuel level even higher by grinding or filing a small recess into the roof of the chamber so that the float can be raised beyond the point where it would normally touch the chamber roof. Do not allow this carb to flood as much as you would an SU. The needle valve must be capable of working at all times. If you have any real problems with surge and can't cure them as above, then try reducing the inside volume of the chamber roof by packing it with plastic metal around the float, just leaving a recess for the float. This should have exactly the same effect as increasing the base volume. (I hope that idea doesn't turn Mr Reece's hair grey.) Frankly there isn't much else you can do to a Reece-Fish as far as gas-flow is concerned.

One of this carb's main advantages is the fact that fuel is dispersed into the airstream as seven small droplets, giving much better atomisation than the one large droplet as on an SU. Also by introducing the fuel at the butterfly one is doing so at the point of greatest gas velocity, again an advantage. Finally other than the butterfly and its spindle there are absolutely no other obstructions to gas-flow.

If I am accused of giving this carb an open plug let me say that I believe in giving credit where credit is due. Not only do these carbs do all that is claimed for them, but Mr Reece is always willing to discuss customer's problems and give advice. Also you can obtain his product on approval, pay if you are satisfied, return it if not. What more can you ask?

NB Since writing, the foregoing modifications have been made to the Reece-Fish 'production line' and different manifold flange fittings are now available to fit almost any type of manifold without modification or the use of adaptors on such cars as the Hillman Imp.

chapter two

Camshaft design and application

Probably one of the most confusing aspects of engine tuning to the average enthusiast, is the apparently large and varied choice of camshafts, many of which are claimed by their vendors to possess new and previously untapped sources of power. Unfortunately as with all sales talk, all too often only virtues are extolled, snags remain unmentioned and many enthusiasts are incapable of recognising that snags do even exist, let alone what they are. This is no fault of the enthusiast for so complicated is the subject of camshaft design and its application that I myself still find some of the finer points a little confusing, such as trying to correlate camshaft design with unfamiliar tuning techniques such as supercharging or true fuel injection.

Fortunately, in this country those, engines that are most popularly tuned such as the BMC A & B series, the Ford in line four-cylinder engine, and the Rootes Imp-based engine, are quite liberally catered for by the more reliable and reputable tuning specialists and to a certain extent the manufacturers themselves. Thus with a certain amount of basic knowledge the enthusiast should, with a little guidance, be able to determine which camshaft best suits his needs.

Camshaft function
As most of you know the camshaft's function in an internal combustion engine is primarily to provide a mechanical means of opening and shutting the inlet and exhaust valves. The valve stem tip or an extension from it (push rod and/or camshaft follower), rests

against the camshaft itself, or more accurately the oil film surrounding the camshaft. The portion of the camshaft in question is not round, but has a distinct profile, like a modified oval or pear shape. The camshaft itself rotates about its own centre axis, thus depending upon what part of the profile the valve stem is touching, it may be closer or further away from this shaft centre. (See diagram.) At X and Y therefore we have the points of fully closed and fully open valves. Points between these represent points of partially open or partially lcosed valves. Usually points X and Y are slightly 'flattened' so that the valve stays fully open or shut for several degrees of camshaft rotation. Thus you can easily see that the further away from the shaft centre that Y is, the greater will be the lift of the valve for any given distance of X from the centre. In actual fact reducing the distance X effectively increases distance Y. This is because, as can be seen from the diagram, the camshaft lifts the valve geometrically by an amount equal to the difference between the distances X and Y, call the lift L. Thus the smaller is X the greater will be L if Y remains constant. Conversely if X is constant L increases if Y is increased. This sounds fine doesn't it?

Increasing a valve's lift does not necessarily decrease engine flexibility. But remember this, there is an obvious physical dimensional restriction on valve lift. Also the greater a valve's lift the greater the distance you have to move a valve in a given time. If a camshaft rotates through 360 degrees, 4000 times a minute, the valve moves a distance equivalent to 2 L (up and down) 4000 times every minute. Increase L by 10 thou and the valve therefore travels 20 thou greater distance every revolution.

Greater stresses
The trouble is that this imposes much more than 20 per cent greater stresses on the valve gear, and hence the need for stronger valve springs (we will deal with this shortly) which themselves pose problems of engine reliability if too strong a valve spring is used. Remember also that the situation is often aggravated by the fact that

tuned engines are usually made to rev more highly than usual and our 4000 rotations per minute become 5000 under normal usage. A further 25 per cent increase in the distance travelled by the valve every minute.

In actual fact as far as the forces acting on the valve gear are concerned, in relation to engine rpm these forces are proportional to the square of the increase in rpm thus doubling engine speed increases the forces by 4 times, and trebling rpm increases the forces 9 times etc.

You will doubtless realise that it is not the actual distance travelled by the valve which is the trouble, other than perhaps its effect on valve guide wear, but the extra speed or acceleration of the valve in travelling these greater distances in a given time. Camshaft designers therefore endeavour to compromise by giving a valve as much lift as possible, while at the same time giving the valve as much time as possible to travel this distance. This is of course expressed in degrees of camshaft rotation.

I mentioned earlier that the points X and Y tend to be 'flattened' to leave the valve open or shut for a short while. This 'dwell' period accounts for a few degrees of camshaft rotation however and still further shortens the time or increases the speed of the valve for any given lift. One must reduce the time which one has to open or shut a valve by half the sum total of the two 'dwell' periods. The diagram should help explain.

As smooth as possible
Unfortunately there is yet one further limiting factor on camshaft profile design, other than its obvious effects on the gas flow characteristics of an engine, which I will deal with later on. One needs to make the profile as smooth as possible. Sudden changes in valve direction such as when a valve starts to close after having been fully open must be avoided. If point Y was literally a point it would be impossible to keep the valve stem in continuous contact with the

camshaft face (the valve's own inertia would carry it on up past point Y) and valve bounce would develop. This is why we use stronger valve springs, to try and make the valve stem stay in contact with the cam face. But stronger springs cause further stresses and consequently there is a limit to what we can use as regards spring strength. Therefore the point at Y must be fairly smooth, as shown in the diagram. These then are the major mechanical factors affecting camshaft profile design.

Large overlaps
You have all heard of camshafts with large overlaps. This is yet another most involved aspect of camshaft design, known as valve overlap. It is really best considered together with gas-flow characteristics of a camshaft. Briefly, overlap is again measured in degrees of camshaft rotation and indicates the length of time that the inlet and exhaust valve on the same cylinder are open *together*. I mentioned that the actual camshaft profile had an influence on gas flow. Profiles are in fact not regular but you will find that the rate of lift imparted to a valve may vary throughout the total lift period of the valve depending upon the degree of camshaft rotation. This profiling is of course related to the position of the piston and to the pressure on the top of the piston at any given position, and often varies from designer to designer. This of course is also related to the amount of overlap and valve acceleration.

Overlap is used basically because the outward travelling exhaust gases from a cylinder tend to create a sucking action helping to draw fresh inlet mixture into the cylinder. At the same time this encouraged flow of inlet gases helps push all burned exhaust gases from the cylinder. Naturally all of this demands that both valves are open together. The longer the time that both valves are open together the greater will be this effect of exhaust gases sucking in new fuel mixture and the fuel mixture pushing out old exhaust gases. Unfortunately once again there is a snag. The exhaust gases need to be travelling very quickly to have any useful effect as also do the inlet gases. This condition does in fact apply on *all* engines even at low engine speeds but only for a very short while. Say we take a normal road engine being driven at 2000rpm. Engine speed is very low and if we were to use a camshaft with a lot of overlap then one would find inlet gases escaping down the exhaust valve and exhaust through the inlet valve. This in fact happens on a racing car and is why there is absolutely no bottom end power or tick-over.

Complete scavenging

However, as engine speed rises the gas speed rises and the back and forward swirling condition of the two gases gradually disappears till the engine comes 'On the cam'. This is the point at which inlet and exhaust gases stop flowing in the wrong directions, no exhaust gases tend to come out of the inlet and everything flows very smoothly. In fact most camshafts on racing engines, are designed to allow the exhaust gases to be completely pushed out by incoming inlet gases, and even allow some inlet mixture to escape down the exhaust just before the latters valve closes. Again this ties in with combustion chamber design. A chamber with lots of nooks and crannies tends to have pockets of gases which cannot easily be exhausted or even ignited. So here again overlap and combustion chamber design must be related at the design stage.

Overlap is also related to exhaust manifold design. A wrongly designed exhaust manifold can prevent the use of a high overlap camshaft, due to the fact that although overlap relates to the amount of time that both inlet and exhaust valves in the *same* cylinder are open, the large overlap also means that several exhaust valves on several *different* cylinders are open at the same time. On a short branch manifold it is possible for exhaust gases to shoot back up an adjacent manifold branch and into a different cylinder. Most upsetting!!

To summarise

Camshaft lift is governed by the speed which is necessarily imparted to a valve to make it travel large distances in short times, and camshaft overlap is governed by the fact that high overlaps ruin engine power at low speeds, which is no good on a road vehicle. As you see camshaft design is extremely complex and specialised; we have only considered the infants school type principles. To go further would mean writing ten times this amount and to cover the subject fully would take several large and complicated volumes full of advanced mathematics which are beyond me anyway.

Buying a camshaft
However, what does the foregoing mean to you and I if we are buying a camshaft? Well, one should consider the following points. First the use of the car. If granny wants to use it for shopping or you are regularly using the car to teach someone to drive then one must use a standard camshaft. Fast road use and rally use do not demand quite such flexible or durable engines and camshafts of greater lift and overlap can be used. Normally neither of the above cases pose such great mechanical problems as an out-and-out full-race 'diabolical' camshafts, and it is this with which I am mostly concerned. The following points apply to camshafts but not to such an extent as with full race cams.

Many tuning specialists offer reground camshafts. These are in fact often standard camshafts reground to a 'different shape'. You will of course realise that to regrind a camshaft one must remove metal. Unfortunately all too often this results in a most diabolical profile causing tremendous valve acceleration and often 'sharp angles' necessitating very, very strong valve springs. Personally I do not like most regrinds that I have seen and at risk of raising a storm, because of course there are notable exceptions, I say avoid regrinds from standard camshafts if you are able to obtain purpose-made specials such as the BMC 731 cam.

Many standard camshafts do not have enough meat in them to be changed from mild to diabolical specifications and still give reliability. On the other hand camshafts which are purpose made or reground

from 'meaty' blanks to even extreme specifications are a good buy.
For example even the old BMC 544, manufactured as such, was a
really fabulous cam. On the other hand regrinds from more standard
cams to the 544 'specification' were the cause of many broken valves,
stems and pitting of camshafts faces etc; the same can be said of the
later BLMC 649 cam.

AUTHOR'S NOTE: In discussing the relative rpm of crankshaft and
camshaft I have for the sake of simplicity suggested that both rotate
at the same speed. In point of fact this is not strictly true.
Camshafts usually rotate at half crankshaft speed, the speed reduction
being facilitated by having different size camshaft and crankshaft
timing wheels (sprockets). Nonetheless this does not alter one bit the
principle and theories discussed.

Extra strong valve springs
I have stated in the foregoing that the use of extra strong valve springs
poses certain problems regarding reliability. What are these problems ?
Well, as you will realise the action of a valve spring is to exert a pull
along the length of the valve stem. Obviously the stronger the spring
the stronger the pull. Thus you can see that theoretically if a strong
enough spring were used then one could stretch the valve stem like
elastic or pull it in halves as one would a plug of plasticine, depending
upon the temper of the valve stem, or otherwise since the valve is
located by the valve head against the cylinder head which acts as a
stop, the valve head could pull right through the cylinder head or
break off by turning inside out like an umbrella.

Of course this does not normally happen, and 'never' happens when
an engine is stationary. However when an engine is turning over at
say, 8000rpm these forces exerted by the spring on the valve are
greatly increased and failure, which is usually in the region of the
valve spring locating collar, occurs. On the other hand even if the
valve is strong enough to withstand these forces, one must remember
that to open a valve the camshaft must push against the spring

strength. Obviously the greater the pressure on the camshaft the greater the wear in this region and the pressure increases as valve spring strength increases. The results are excessively rapid wear or even pitting of the camshaft face or breakage of a camshaft follower.

Besides choosing a camshaft of good design how best can we reduce the necessity for terribly strong springs? Let us consider the effect of reducing the weight of the valves, camshaft followers and push rods.

Reducing the strength
I have stated before that this is an excellent way of preventing valve bounce or reducing the strength of the spring which is needed to prevent bounce. But let us now consider why this is so.

Let us consider an earlier statement that I made—'If Y was literally a point then the valve's own inertia would carry it on up past this point'. If you tie a ½oz weight to a piece of cord and project it from, say, a schoolboy's catapult it will travel at say 100mph till the cord becomes tight and then stop. If on the other hand we tie a ½lb weight to the same cord and project it from the catapult at the same speed then the cord will snap as soon as it becomes tight. This is common sense and we say that it is because of the greater momentum of the heavier weight.

We can relate the valve and other 'rotating' parts of the valve gear to our ½oz or ½lb weights and our valve spring to our cord. If we use a ½lb weight and a much stronger cord, the latter will not break. Similarly if we use a heavy valve and a very strong spring then at say 8000rpm (maximum rpm on the engine concerned) the spring will be strong enough to prevent our heavy valve, which is travelling at maximum speed, passing on beyond point Y. On the other hand if we reduce spring strength the valve inertia at its maximum speed will be such that it will carry on up past Y into 'mid-air', since the spring cannot control such inertia in such a short space of time, caused by sudden change in direction, especially if Y is a sharp point. This is the point of valve bounce and is known as such. However, if we increase the strength of our valve spring to a point where it is capable of dealing with the valve's inertia at maximum speed we find that it will keep the valve pressed down on to the camshaft face even if Y is a sharp point, and valve bounce will not occur.

Unfortunately the spring strength necessary to do this may be such as to cause unreliability in the directions mentioned earlier.
If however, we reduce the weight of the valve and other reciprocating parts we effectively reduce the inertia (ie. our $\frac{1}{2}$lb weight becomes $\frac{1}{2}$oz) and a much weaker spring is capable of controlling the inertia of maximum valve gear speed. The weaker spring also allows for greater reliability. The danger of valve bounce is of course that the valve carries on beyond point Y down towards the piston and the whole engine carries on rotating. The piston comes up to tdc at which stage one would expect the valve to be far enough away to avoid contact. If valve bounce occurs the valve is further open than it should be and hence is quite liable to strike the piston, with obvious disastrous results, or the valve on being pulled back into the camshaft face strikes it at far greater speed than it should do, the shock of which can itself cause breakage.

At school one learns that the smallest particle capable of individual existence is the atom. Any graduate who has studied atomic theory will tell you that this is an over-simplification of the facts. The same applies to my explanation on camshafts. But the basic information is correct and for all practical purposes is all the knowledge one needs unless one is intent on designing one's own cam. In this case I would say you must learn for yourself—I cannot teach you anyway. At the beginning of this chapter I attempted to throw a glimmer of light on the more obvious problems associated with camshaft design, with particular regard to the racing engine. I will now try to throw a little more light on early camshafts. These problems are best explained if we first consider the simple or old fashioned principles of camshaft design, which are carried out with regard only to geometrical limitations.

Early camshafts consisted in profile of three separate radii or arcs. One had the base circle, the flank radius and the nose radius. As we can see from the diagram, all the time the valve, or its extension, is in contact with the base circle then it remains stationary, neither opening nor closing. As it contacts the flank radius it starts to open, suddenly and with constant acceleration. Then the small nose radius is reached and the valve starts to slow down finally coming to rest on the nose. At first this may all seem just fine, the profile can be based on past experience and suited to current needs according to certain preconceived rules. It is this reliance on past experience that has done so much to give the layman the idea that successful camshaft designing depends upon luck and inspiration drawn either from Holy quarters or the Devil, depending upon his opinion of a particular camshaft in use at the time.

Unfortunately these ideas break down in many respects though they are still used by some so-called tuners. Camshafts designed by early methods really show their inadequacies at the very high rpm such as used on most modern racing units. Even at very low revs early cams resulted in noisy and harsh valve gear and, though this could be tolerated on a racing engine, as rev limits have risen the older cams have caused valve-gear breakage, regardless of the simple geometric changes that were made.

The basic problem was that designers started designing with completely the wrong approach. Cams were designed with little regard to the forces which they generated in the valve gear. They should be designed after first studying the forces which would be generated. The more enlightened did, and still do this and it was a big step forward, but was still too elementary in that one vital factor was ignored; valve gear flex. Theories were all based on the assumption that the valve gear was rigid. How can it be when to start with you have a stationary tappet gap of say 12 thou? Further, under heavy loads such as experienced using very strong springs and 'diabolical' cams, flexing occurs, especially with pushrod set ups. Camshafts and rocker shafts whip, push rods and rockers bend and flex, valves wobble and fixing studs and pedestals bend and flex.

Wear on the valve gear

I have explained why it was necessary to use strong valve springs if using high rpm and how this was accentuated with heavy valve gear and a diabolical cam.

At low engine speeds, say 4000rpm, the forces imparted to the valve are not very great, the valve's acceleration being quite low and the springs keep the valve pressed hard down on the camshaft face with considerable force. This force is greater at the sharpest point and hence it is the nose radius which experiences the greater wear and racing engines which are habitually under revved often show severe pitting at the nose radius. The reason for the forces being greater at the nose is that being a small radius only a small part is in contact the valve or its extension, the cam follower, at any one time and hence the full force of the strong valve spring is transmitted through a small point. This is the same factor which causes the 10 stone woman's stiletto heels to make dents in the floor whilst 15 stone man's shoe does not even leave a mark. Also the valve spring is fully compressed at the nose. On the other hand at peak rpm as explained earlier the valve is tending to bounce or 'float' and this is particularly noticed at the sharp nose, and hence this point tends to be more lightly loaded and experiences less wear. At the same time it is the flank radius which is actually pushing the valve open and since this is a very large flat radius, one finds a very heavy weight, but spread over a large area. Again wear is consequently not too much of a problem.

Let's now consider the point of valve bounce. The valve fails to touch the cam nose and floats on past it. Eventually it is stopped and immediately whipped back into contact with the cam, bounces off again and, unless engine rpm is reduced, may continue to bounce for the whole 360° rotation of the camshaft. Fortunately the initial greater landing shock is usually taken by the flank radius. Due to the camshaft's continued rotation the load is spread and with 'luck' the cam suffers few ill effects. On the other hand terrific shock impulses are transmitted to the reciprocating valve gear such as push rods, valves, springs, collars, sprockets and chains, and breakage often occurs. These forces or shock impulses arise from the continued *sudden* changes in direction (negative forces becoming positive forces and vice-versa) of the valve gear.

Tappet clearance
Before I really get carried away and confuse myself I would like to consider one other problem. The static tappet clearance.

As can be seen from old fashioned camshafts, all the time the valve is closed, with the camshaft follower and or push rod are resting on the base circle, a fixed gap of say 15 thou exists. Suddenly the flank radius is reached and the follower and push rod are rapidly accelerated to maximum speed. At this speed it strikes the rocker arm with a tremendous hammer blow and, or the rocker strikes the valve an equally heavy blow. Forces exerted on the valve gear and camshaft are momentarily enormous and everything bends or flexes. Like an elastic band with a weight on it, it straightens up but rebounds and continues to flex on a diminishing scale. Again a factor causing unreliability. These flexings naturally are transmitted directly to the valve springs and cause them to 'surge'. This is understandable when you consider that it is the valve spring which is endeavouring to dampen down or prevent the initial and secondary 'rebound flexing'.

Whilst it is not possible to consider modern camshaft designing in detail within these pages I can tell you that much of the work is so complicated that the normal human mind is almost unable to cope, at least in a reasonable amount of time. Consequently many modern cams are 'computer designed', the Americans probably being the real pioneers of such techniques. 'Computerised', or computer designed cams often reduce the tendency for valve bounce by so designing the profile that the valve is slowed down just before the point of full lift is reached. Ramps are also included so that initial uptake of the tappet clearance is gradual and steady, valve gear is not suddenly accelerated and thus the initial hammer blow impact is avoided, all of which greatly increases valve gear reliability.

Such cams are often known as polydyne camshafts and account for the fact that in the past engineers in the USA have been able to extract fantastic power quite reliably from apparently old fashioned push-rod engines. The term 'polydyne' is derived from the type of curve obtained when we plot graphically on paper the forces imparted to the valve gear, the rate of lift of the camshaft, degrees of camshaft rotation and rpm. As I have said before there are many very advanced volumes written by brilliant mathematicians on this subject, and if curiosity leads you onwards read them, if you can. I never intend to design a polydyne camshaft, but I do like to appreciate the problems associated with camshaft design.

BLMC 'A' series camshafts

Fortunately those people wishing to tune a BLMC A Series Engine have a very wide choice of easily obtainable camshafts, marketed by BLMC themselves, some of which were, or still, are fitted to normal standard production vehicles. More of these in a moment, however. First, a brief word on regrinds for this particular engine.

Avoid any supposed race specification regrind which is not based on the C-AEA648 (649) camshaft. This, providing it is done properly, can be slightly modified to a different specification. Normally, little metal need be removed as it already has a fair amount of lift and overlap, neither of which can be radically increased. Further, any regrind camshaft should have been specially treated to harden it, either by nitriding or tuftriding it. Even this however has its problems in that it can cause shaft distortion and any treated regrind must be carefully checked for this.

It is only when after the last $\frac{1}{2}$ bhp or so on a racing engine that one need even contemplate modifying a BLMC camshaft and most people still use an ordinary C-AEA648 cam for racing. Consider first which vehicles utilise the BLMC A series engines. In order of engine capacity we have the following list.

848cc Ordinary Minis, early Elfs and Hornets and latterly A35 vans.
948cc A35, Farina A40, Minor 1000, Sprite Mk I and Mk II, Midget Mk I.
970cc Cooper S.
997cc Early Cooper Minis.
998cc Late Coopers, Elfs and Hornets.
1071cc Early Cooper S.
1098cc Farina A40, Minor 1000, Sprite Mk II, Sprite Mk III, Midget Mk I and Mk II, Morris and Austin, Riley and Wolseley, MG 1100's.
1275cc Cooper S and latest Sprites and Midgets and 1300 units.

Basically one can use the latest BLMC camshaft in any one of these vehicles with somewhat similar effects bearing in mind that the camshaft must be capable of giving useable power within the safe rev range of each particular engine. Perhaps I should first run through these various camshafts and give a broad outline of their uses and then consider each type of A series engine and say what can and cannot be used. Remember also that the smaller the engine capacity the more it will tend to lose useable bottom end power with a diabolical camshaft.

AEA630 Really only worth fitting on the 848 Mini as a fairly mild road camshaft. Gives excellent bottom end power even on 848 Minis, but must not be taken over 6000 to 6200rpm as it is very dangerous to the inlet valve gear due to the tremendous accelerations imparted to the inlet valve. Little top end power.

AEG148 Similar to AEA630 in most respects.

88G229 Formerly known as 2A948. A mild rally camshaft which works quite well on all engines but does destroy bottom end power on 848 Minis and runs out of power at about 6500rpm. It has the advantage of being cheap and easy to obtain and is not too harsh on valve gear.

AEG510 I have little experience of this cam but I would say that one can expect results very similar to those obtained with the 88G229 camshaft.

Anyway, the effects of the camshafts mentioned above are pretty marginal and many people don't even bother to change to one of these preferring to keep the standard article. The 88G229 is the most widely used and regarded by most as the old faithful. Following on however, we come to the 'diabolical' camshafts.

C-AEA731 This really is a rather fabulous 'Jack of all Trades'. It can be regarded as a rather vigorous rally or a semi-race camshaft. It is also a must for the really keen road-man, who is prepared to put up with little tick over or very low down power. However, if an engine is properly tuned it gives excellent power from 3500rpm or so and goes on giving power to 7000+ and it is not too hard on the valve gear.

C-AEA648 Commonly known by most people as the 649. An out-and-out full race camshaft. No power below 5000 (in some cases perhaps 4-4500) with power in plenty up to 8500+. One cannot use it with normal valves as the lift is so great that spring crush occurs. One must use Cooper S length valves or machine the spring locating grooves away to expose more valve stem and allow greater spring travel. I must emphasis this as it is most important. Frankly, the easiest way is to use Cooper S valves and have the heads turned down to size. This camshaft is very hard on valve gear, especially camshaft followers. If using any of the other cams not originally designed for the Cooper S, in a Cooper S one must modify the oil pump's projecting slotted finger.

The C-AEG529 camshaft is identical to the 649 but has a different oil pump drive layout suited to the oil pump on the latest 1275 Sprites and Midgets and 1300 units.

Just recently BLMC Special Tuning have brought out two new camshafts which are still something of a mystery both as to exact specification and what engines they are suited to. Both have been designed for the fuel injection 1275 Cooper S unit. Both have the same spline drive to the oil pump as on the C-AEG529 just mentioned. Both are reputed to have the same amount of lift as the 529 and 649 but differ in having extended valve periods or overlaps. The milder of the two, C-AEG597 has an inlet period of 320° and an exhaust of 300°. The more extreme, the C-AEG595 has periods of 320° for both inlet and exhaust. Thus both are more extreme than the 649/529 cams which have periods of 300° for both inlet and exhaust. It is true to say that both of the later cams have less bottom end power though it is to be hoped that they will increase an engine's top-end output. Time will tell.

I must stress that all of the out and out racing cams mentioned make very heavy demands on valve gear reliability and not only do they tend to cause spring crush but unless valves themselves are made of best quality racing Nimonic materials they will almost certainly break after a short period of use. I would therefore use nothing other than Cooper S inlet and exhaust, Sprite Mk 4/Mk 3 Midget 1300 exhaust valves etc. as sold by BLMC Special Tuning Department or other reputable tuning company who sell special Nimonic valves.

I will just mention the old 544 camshaft. This is no longer made and has been replaced by the C-AEA648.

Let us now take a look at the various engines.

848cc engine

The maximum safe rpm on this engine are 6000 to 6500 providing one ignores the timing chain lift. (The chain tensioners disintegrate at about 5500). Some people wisely advocate less than this and set 6000 as a regular limit, and these people include myself. Personally I reckon a standard cam is OK for road use or an AEA630 for rallies, but on the other hand if one carefully builds up an 848 engine to racing standards then one can use 7500 and very occasionally 8000 rpm with *reasonable* safety, which enables one to use the C-AEA648 camshaft with suitable valves and springs. Such an engine would be useless on the road and if road use is envisaged you must not go beyond the 731 camshaft. By racing standards I mean, balance the engine and fit a crankshaft vibration damper and the latest crankshaft.

Remember this however, changing a camshaft on this engine necessitates a complete engine strip down, line boring of the camshaft housing and the fitting of white metal camshaft bearings. If you don't do this and allow the cam to run direct in the block, as standard cams do, you will *certainly* seize up. Be warned. The same also applies if fitting very strong valve springs capable of over 6500 or so.

948cc engine
I personally would consider the engine in exactly the same light as the 848 engine with the exception that I would use the 88G229 camshaft as my mildest improvement instead of the AEA630. This engine also needs to have the camshaft housing line bored and fitted with white metal bearings.

997cc engine
Used only in the early Coopers. Fully prepared capable of a fairly reliable 7200-7500. This was the first engine fitted with camshaft bearings as standard. The 2A948 was standard fitting. I would not use the C-AEA648 cam, limiting myself to the 731 or if I could obtain one, the older 544 camshaft. I say this not because in standard form this engine is any less reliable than the 848 or 948cc engines, in fact quite the reverse, but the truth of the matter is that this is now an obsolete engine and few people cater for it, which means that these days it cannot be developed to the same extent as the current engines.

998cc engine
Similar but somewhat better than the 997 engine, having a better bore to stroke ratio and rather better reliability. The 998 Cooper uses the AEA630 cam as standard. Thus the 88G229 is an excellent road/rally improvement or even better one can use the 731 cam, bearing in mind my earlier comment. Some people still race this engine or derivatives from it. Thus, bearing in mind a safe limit of 7500 with occasional

8000, the C-AEA648 camshaft is an excellent fitting, bearing in mind my earlier comments on the necessity for the correct valves and springs.

1098cc engine

This engine comes in two forms, the normal and that fitted to the Mk III Sprites and Mk II Midget. The latter two being fitted with much more rigid blocks and sporting 2in diameter crankshaft main bearing journals. The more normal is fitted to transverse 1100s and Mk I and II Sprites, Mk I Midgets, etc.

Maximum rpm should be kept below 6500 on this rather woolly and low revving engine, I would never suggest anything more vigorous than a 88G229 camshaft, because if you do use, say a 731 or 649 and take advantage of the extra power, then as sure as I sit here there will be one almighty *bang !* Then you will be much poorer. However the 1098 engine fitted with the heavier crankshaft is a rather different proposition providing one fits the special BLMC competition crankshaft etc—no goodies are offered for the former engine. Under the circumstances this 1098 engine becomes very reliable and any camshaft is OK, though I would limit my choice to the 731 or the 649.

970, 1071 and 1275cc engines

I have lumped all these engines together because basically one can consider them as being identical from a camshaft tuning aspect. The 731 makes an excellent general purpose camshaft and the 649 cam was designed for the Cooper S anyway.

The 1275cc Sprite Midget and 1300 engines cannot use any of the aforementioned camshafts other than those such as the C-AEG529 (similar to 648), G-AEG542 (similar to 731), C-AEG595, C-AEG597.

To summarise.

The C-AEA648 designed as a full race camshaft for the Cooper S which when fitted to any other A series engine (cannot be fitted to

A Series camshaft details.

PART NOS.	8G 712 / 2A 297 / 2A 571 (Pin drive oil pump)	12G 165 / AEA 630	AEG 148	88G 229 / 2A 948	AEG 510	C-AEA 731	C-AEA 544	C-AEA 648	—	—
MARKING		2 rings	—	1 ring	1 ring	3 rings	AEA 544	AEA 649	—	—
STANDARD USE	Mini 850 / Mini 998 / 9C Sprite 1 / 9C GAH/MG	1100 range / 998 Cooper / 9CGAH/MG	Early Cooper 'S' / 10CC AH/MG	997 Cooper	Cooper 'S' From: 9FSAY-40006	Race	Early full race	Race	—	—
CAM LOBE WIDTH	3/8"	3/8"	1/2"	3/8"	1/2"	3/8"	1/2"	1/2"	—	—
INLET OPENS B.T.D.C.	5°	5°	5°	16°	10°	24°	34°	50°	60°	60°
INLET CLOSES A.B.D.C.	45°	45°	45°	56°	50°	64°	74°	70°	80°	80°
EXHAUST OPENS B.B.D.C.	40°	51°	51°	51°	51°	59°	69°	75°	75°	85°
EXHAUST CLOSES A.T.D.C.	10°	21°	21°	21°	21°	29°	39°	45°	45°	55°
INLET PERIOD	230°	230°	230°	252°	240°	268°	288°	300°	320°	320°
EXHAUST PERIOD	230°	252°	252°	252°	252°	268°	288°	300°	300°	320°
CAM LIFT	.221"	.250"	.250"	.250"	.250"	.252"	.306"	.315"	.315"	.315"
VALVE LIFT	.285"	.318"	.318"	.318"	.318"	.320"	.388"	.394"	.394"	.394"
RUNNING CLEARANCE	.012"	.012"	.012"	.015"	.015"	.015"	.015"	.015"	.015"	.015"
TIMING CLEARANCE	.019"	.019"	.021"	.019"	.021"	.021"	.021"	.021"	.021"	.021"

PART NO.	12A 1065 (Spider drive *oil pump)	12G 726	AEG 522 / AEG 537	C-AEG 567	C-AEG 542	C-AEG 529	C-AEG 597	C-AEG 595
MARKING		2 rings	—	AEG 567	1 ring / AEG 543	AEG 530	AEG 598	AEE 596
STANDARD USE	Automatic Mini 850/998	Automatic 1100	12CC AH/MG Semi-race	Semi-race	Rally Road	Race	Race	Race
CAM LOBE WIDTH	3/8"	3/8"	1/2"	1/2"	1/2"	1/2"	1/2"	1/2"

*Requires flange 12G 729

latest Sprites and Midgets) must have provision made by the use of longer valves and springs or equivalent, to accommodate its extreme lift.

When using these racing camshafts one may need to use longer tappet adjusting screws ($1\frac{1}{2}$in) or machine 50 thous from the rocker shaft pedestal bottoms, to obtain the correct rocker to valve stem tip angularity. Further if one uses the extra strong Cooper S valve springs one needs to use the special bottom locating cups supplied by BLMC Special Tuning department.

chapter three

Lightening valve gear

The relative merits and demerits have often been discussed by myself
and others in the past, and many thousands of words have been
written explaining how, by reducing valve gear weight one can raise
the point of valve bounce on an engine, or alternatively reduce the
strength of valve spring needed for a particular valve bounce point.
In this chapter I intend using the BLMC A-series engine as an
example to explain what to lighten, where to lighten it and how to do
the work. Though aimed at the A series unit the following applies to
almost any push-rod engine, and indeed to any reciprocating engine
with valves.

Starting at the camshaft end, we find the cylindrical or bucket-shaped
camshaft followers or tappet blocks. These are specially case hardened
at manufacture and this makes their modification by ordinary hand
tools almost impossible. I always take mine to a good machinist who
modifies them to my specification on a lathe, at a cost of £1 10s per set.
I have the overall length reduced by $\frac{1}{8}$in. The inside is then machined
out to reduce the wall thickness almost to eggshell dimensions at the
top open end, at the same time ensuring that the bottom has a nicely
blended radius into the base, to minimise the possibility of the base
breaking away from the side walls. These machining operations do,
in fact, have a secondary advantage.
It will be found that approximately 20 per cent of all BLMC A series
camshaft followers are unsuitable for use in a racing engine, being
over-hardened and too brittle. This is almost impossible to discern

until one attempts machining. Those that are too hard cannot be machined, they just crack or break the cutting tool. They must be thrown away. Thus to make up a full set of nine followers one being a spare, you will probably need 12 followers. Don't send the over-hardened ones back to BLMC on warranty. They are quite alright for normal use, and you will just get laughed at. They are not designed for racing use, or at least their final inspection process isn't, just normal road use.

In one of the early editions of my book—'Tuning a Mini', I suggested that one could machine a flat on the outside of tappet blocks to increase oil splash and lubrication of the valve gear. This was based on the tappet blocks found in a 'works' engine purchased by myself in the early days of Mini tuning. This *must not* be done. Tappet blocks rotate and since the size and shape of their base determines to a large extent the opening period of the valve, one can see that an irregularly-shaped base will cause an irregular valve period. Frankly, I am amazed, looking back, that this had not been realised by such experienced engine tuners as produced this engine, it just shows that even the experts don't know everything.

One often sees adverts claiming outstanding properties of lightness and strength for alloy push-rods. Not only do they lack the strength and rigidity of their steel counterparts, size for size, but the rate of expansion due to temperature increase is so great and shows such variations according to temperature, that they are worse than useless.

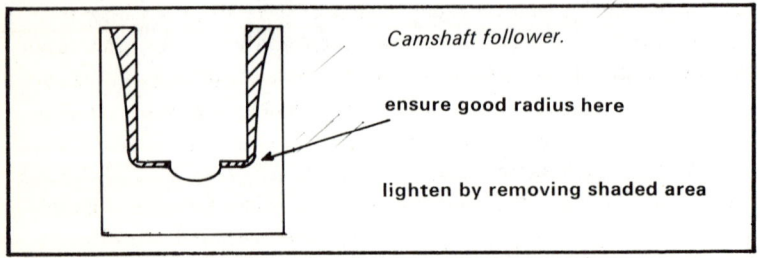

Camshaft follower.

ensure good radius here

lighten by removing shaded area

Tappet settings become meaningless! There is no substitute for steel, save perhaps titanium, which is very, very expensive. Some people have successfully used tubular steel pushrods and although I was once a great believer in them, I am not so sure these days and I feel that they lack rigidity and increase valve-gear flexing which is most detrimental. No, these days I stick to the standard article, lightened.

The only places where metal can be removed are at the extreme ends, from the foot and top cup. This latter can be machined on the outside to reduce the overall size and thickness. On no account must you remove metal from the inside of the cup. The foot can be reduced by removing metal from the uppermost portion. Metal must not be removed from the sole of the foot.

This work can all be done free-hand using an ordinary bench-mounted grindstone or large abrasive disc, but it is rather tedious and considerable care is necessary if one is to avoid 'necking' the rod at the points where the cup and foot join the shank. If this happens, it greatly weakens the rod and necessitates its replacement. Better that this operation should be performed on a lathe as it is very simple and hence cheap. I have done it free-hand, but needed several spare rods to replace the 'necked' components.

Most people associate valve-gear lightening with the removal of metal from the rockers alone, often ignoring all other parts of the valve gear. Although it is advantageous to reduce weight in all reciprocating parts of the valve gear, it is true to say that often the greatest savings in weight are made in the rockers, providing they have suitable reserves of strength and are of the cast and not pressed steel, type. The latter should be replaced wherever possible. Often one can find that different models, using basically similar engines, utilise different rockers which are identical with regard to ratio of leverages, etc., but which are often much lighter. The latest Cooper S rockers AEG 425 are the perfect example of this.

The most effective savings in rocker weight are in those areas which travel the greatest distance and reach the highest speeds or are subject to the greatest acceleration, the extreme ends. Fortunately these are also the areas which are least subject to breakage.

Taking first the valve stem end. Usually this is just a solid rectangular lump of metal, obviously far heavier than need be, with the excess material in the width of the pad (as I call this lump) not the depth. This excess width is to compensate for any inaccuracies in rocker alignment over the valve stems. Since the pad is so wide, no matter how inaccurate the alignment, some part is bound to be over the valve stem and able to push the valve open. If one knows that the rockers align accurately with the valves then one can reduce the pad width almost to the diameter of the valve stem. If, on the other hand, alignment is uncertain or inaccurate, then the bottom of the pad which presses on the valve must not be reduced in width, but one can machine the pad to a wedge shape.

On no account must metal be removed from the sole of the pad on rockers. This is machined to a set radius and must not be altered. This radius is set so that although the rocker movement is an arc of a circle, the pressing action on the valve stem remains vertical, in line with the length of the valve stem, not putting any side thrust forces on to the valve, which would tend to bend it and wear guides out.

At the adjuster end one often finds a large lump of weld and, or ridges left in the casting. Obviously these should both be removed. Sometimes removal of the weld lump uncovers an oil-way drilling which it was intended to blank off. In my opinion it is relatively

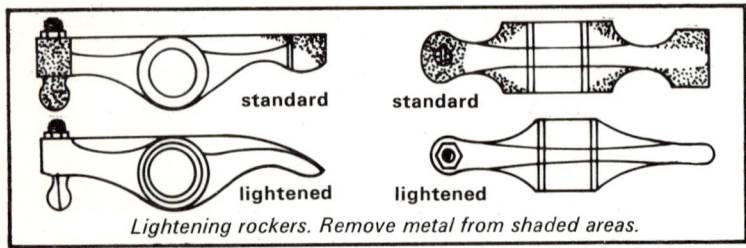

standard   standard
lightened   lightened

Lightening rockers. Remove metal from shaded areas.

unimportant whether this hole be bunged up with a little solder or plastic metal or left open.

The rounded section of the rocker which houses the adjuster and is threaded internally, can have its outside diameter reduced by an amount which will vary according to the original thickness of the metal used and your own commonsense. I usually make the outside diameter approximately the same as that of the locking nut used on the screw adjuster. The bottom edges of the rounded section can be bevelled off, but on no account must metal be taken from the top where the locking nut seats, otherwise the tappets work loose.

You will often find that the screw adjuster has a very deep screwdriver slot in it, the depth of which can be greatly reduced by removing metal from the screw adjuster. The locking nuts often have part numbers or trade numbers embossed on them. These can be machined off for very small savings. It is possible to reduce the ball of the screw adjuster to a pear shape, but since it is a very hard material a precision grinder is needed for best results. I don't bother with this. All other work can be carried out with simple hand tools and or bench grinders or abrasive discs.

I have previously suggested that one could polish and lighten the whole of a rocker and while it is true that very small gains are to be had, they are very small and it is doubtful whether the effects could be any better than a rocker which has been properly lightened at the extreme ends. Such intricate polishing and finishing is time absorbing.

One word of warning: you must remember that rockers suffer very heavy stresses and in my opinion no attempt should be made to reduce a rocker's depth at *any* point other than the *extreme* edges of the ends, which can be bevelled off, as breakage is almost impossible at these points. A rocker relies on its depth for its strength, and strength should never be sacrificed for lightness without very careful and expert consideration or expensive trial and error.

It is not really possible to lighten BLMC spring collars and collets and retain adequate strength. On the other hand I fancy that the use of dural collars, as often used on motorcycles, would show very worthwhile savings if you could get them made. I only wish I had the guts to try them, but oh boy! What a disaster if they broke at 8500! Considerable savings can be made on the valves when they are reshaped to improve gas-flow.

Usually the camshaft mounted timing chain sprocket can be considerably lightened by machining and drilling, provided that it is the steel type such as fitted to the Cooper S. However, this merely reduces the overall weight of the car and the load on the timing chain and camshaft. It has no influence on valve bounce or spring strength.

Considerations of Valve Shape

A few moments ago I mentioned reshaping of valves to reduce weight and improve gas flow. Although I have explained the principles and shape required before (see 'Tuning a Mini'), I did not say why this shape was best and I am often asked why a true tulip-shaped valve is not ideal, and why we use a penny-on-a-stick type of valve in BLMC engines.

Basically, the reason is this. Tulip valves work very well on those cylinder heads which have a straight port into the combustion chamber, such as downdraught port, especially if the combustion chamber is hemispherical with no combustion chamber wall close to the edge of a valve. On the BLMC engine, however, we find a side-draught port with a very sharp right angle bend in it, then a short run-in to the combustion chamber known as the valve throat. The edge of the valve head comes very close to the combustion chamber wall. Unfortunately, these characteristics so alter the gas-flow that a true tulip valve is 'useless' and a penny-on-a-stick is needed. Let us look at it from another angle—the purely physical angle.

As can be seen from the diagram, at full lift the shape of a tulip valve is such that the clear area around the valve seat is far less than with a 'penny-on-a-stick', and this effect is even more noticeable at, say, half-lift. Thus by reducing the clear area it in fact reduces the gap between the valve and the head seat through which the gases can escape. This sets up a reverse pressure to the ingoing fuel mixture, slows them down, and reduces the total amount of fuel mixture which it is possible to get into a cylinder. Exactly the reverse of what we are aiming at when tuning an engine. With a BLMC engine, ingoing gases tend to spread into the cylinder under and around the valve. This is not quite the same as with a down-draught port feeding into a hemispherical head, where the gases tend to shoot into the cylinder straight past the valve.

In the former case, we need to leave as much clear space for the gases to spread past valves, as freely and quickly as possible, hence the 'penny-on-a-stick' and the large distance between the underside of the valve and the head seat. Where the gases shoot straight into a cylinder, the valve needs to offer as little resistance to the shooting action as possible hence the streamlined shape resembling a tulip head. Needless to say, the 'penny-on-a-stick' is a lot lighter than a tulip valve, though the latter sometimes have hollowed out heads to reduce their weight. Unfortunately, this not only lowers the compression ratio, which would be unacceptable on something like an 850 Mini, but also makes a nasty little pocket in the combustion chamber in which unburned gases can collect.

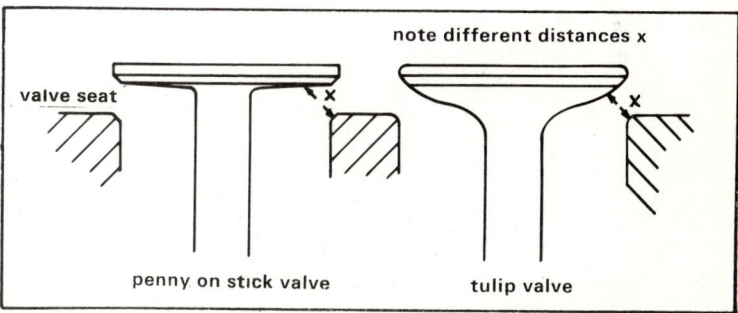

note different distances x

valve seat

x

x

penny on stick valve

tulip valve

A further aid to lightening valves where the head has a large heavy mass such as on Tulip valves, is to fill them with sodium, but such valves are very expensive and rare other than on the most exotic machinery. Sodium was, and indeed still is, used to fill some of the larger exhaust valves, not only to reduce weight but to improve the cooling of the valve.

chapter four

Changing oil pump types

Until the inception of the 1275 Sprite/Midget, BLMC A-series
engines were all lubricated by means of oil pumps driven off the
camshaft via a slotted shaft/cam-mounted pin arrangement.

The 1275 Sprite/Midget, and later the 1300 non-Cooper S units,
sported somewhat similar pumps, but these were, and still are driven
not via the slotted shaft/pin arrangement, but through a splined shaft
which fitted into an internally splined washer, located within the
tail of the cam. This later set-up was generally considered more
reliable than the early type, in that there was no tendency for the
oil-pump rotor shaft to shear when engines were started from cold,
this problem being prevalent on those engines developing 100psi
oil pressure and, at the same time, using thick castor-based vegetable
oils. However, due to problems involved in the fitting of the spider
pump, most people continued to use the earlier pin-drive set-up.

It is, I think, worth pointing out at this stage that to date there are
two basic pump types which are of interest to the enthusiast. These
are the spider-drive variety which, for brevity I will call the spider
pumps and the earlier pin/slotted shaft type, which I will call the pin
pump. Pin pumps are themselves divided into two types, one being
used on Cooper S, the other on all other non-1300 A series units.
The difference lies in the fact that the S pump has a longer rotor
shaft the one that is slotted to take the drive to compensate for the
extra length of the S block measured between the front face and inner

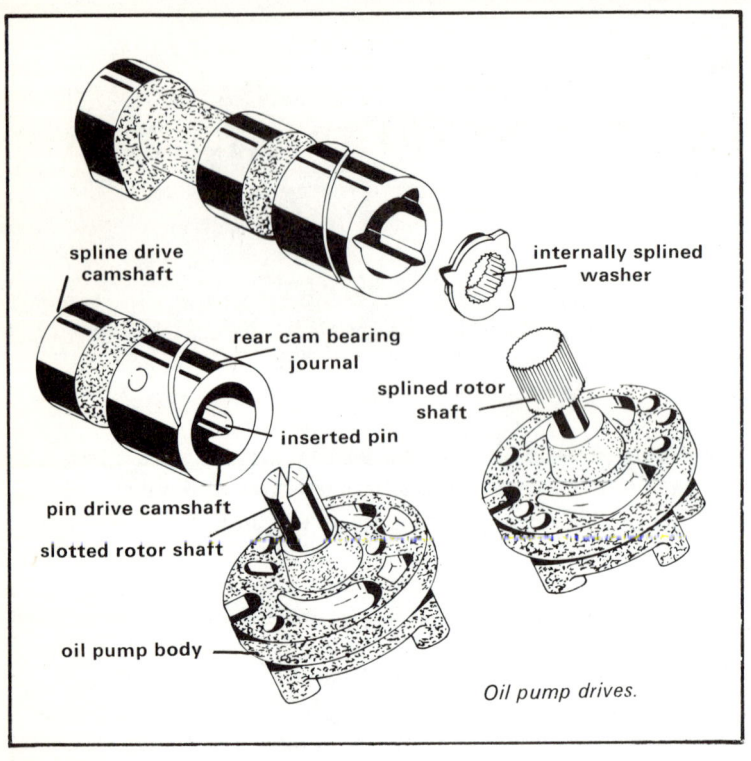

spline drive camshaft

internally splined washer

rear cam bearing journal

splined rotor shaft

inserted pin

pin drive camshaft

slotted rotor shaft

oil pump body

Oil pump drives.

face of the pump housing, the S being some 225 thou longer. This feature of extra length is shared with the 1300 blocks.

So much for background, what about fitting the splined pump to non-1300 units ? Well at first two problems are evident :—

(1) The camshafts which drive the splined-pump differ from those used on pin-drive set-ups. Instead of a pin being inserted into the tail of the cam, the cam tail is shaped to accept the location of an internally splined 'washer' within it. The two types are *not* interchangeable.

(2) The pin-pump is retained in the block by three bolts whereas the splined-pump fixes via four similar but shorter bolts.

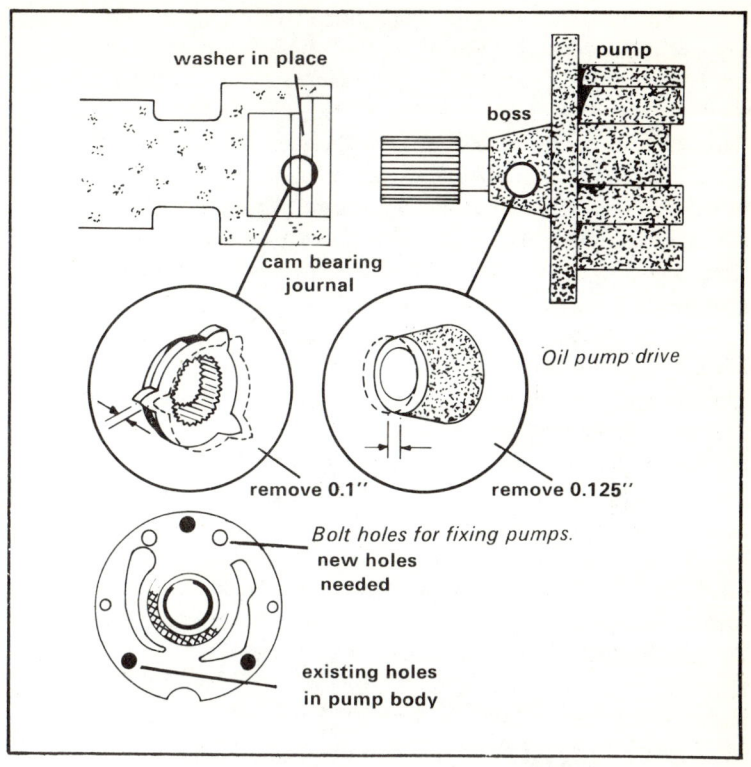

washer in place

pump

boss

cam bearing
journal

Oil pump drive

remove 0.1″

remove 0.125″

Bolt holes for fixing pumps.
new holes
needed

existing holes
in pump body

As far as the Cooper S is concerned, all that is necessary to fit the
later pump is to drill two extra holes into the block and tap suitable
threads into them to accept the extra bolts. Also, it is of course
necessary to fit the appropriate later type camshaft. With these
conditions met, it merely becomes a case of bolting the whole lot
together. With the introduction of two new BLMC camshafts, the
C-AEG595 and C-AEG597, both of which show increased valve
opening periods when compared to the 649 cam, the fitting of the
later pumps has become quite common since both of the new cams
are designed for spider drive.

53

It is when one comes to apply this spline-drive set-up to other A-series engines such as the 850 Mini or the 998 that the real problems arise, as I found out when I came to fit this arrangement to my 850 racer. As is the case with the Cooper S, two extra tapped holes were needed in the block pump housing. When I then came to fit the whole lot together, I found that, fully located, the camshaft-nose protruded from the front of the block some 275 thou, which is 225 thou more than it should be. I had in fact expected this since the pump was designed for the 1300, or suitably modified S unit and accounted for the extra block length on those units.

What was happening was that the splined pump shaft was sliding right through the cam-mounted splined 'washer' and allowing the shaft boss, which is an integral part of the pump body, to foul against the 'washer' and so push the cam forward in the block. It was not possible just to grind the end off the cam nose, assuming one had the necessary equipment for such a job, because, in pushing the cam forward, the oil-way grooves in the cam journals no longer lined up with the cam bearing oilways in the block. To run an engine thus, would surely cause a major blow-up through oil starvation. I decided that the only way out was to remove metal from the 'washer' and from the pump boss. Accordingly, I ground 100 thou off the 'washer'—making it thinner, and took 125 thou off the end of the pump boss—making it shorter. On reassembly, I found that the cam now only stuck out at the front of the block by the required 50 thou.

However, another problem manifested itself. Although the pump boss no longer fouled the 'washer' , I found that the splines in the latter no longer mated up with those on the pump shaft. The shaft had passed right through the 'washer' and contact was only being made over about $\frac{1}{32}$in. In use, this would cause the splines to shear and result in the pump producing zero oil pressure. Fortunately, for this particular application, what is in my opinion a very serious design fault enabled me to overcome this problem.

The splined shaft is merely a simple press-fit within the pump rotor

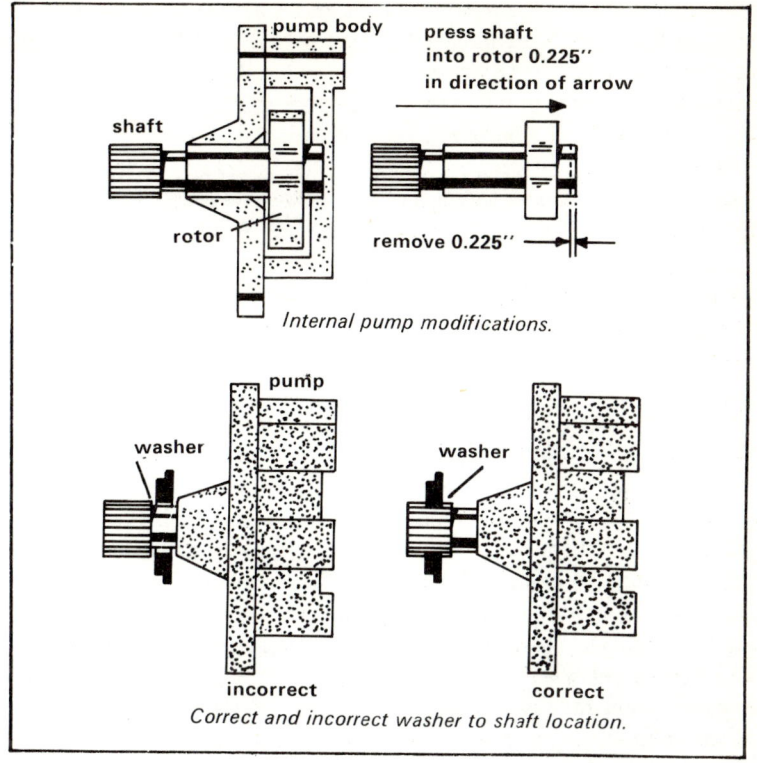

pump body

press shaft
into rotor 0.225″
in direction of arrow

shaft

rotor

remove 0.225″ →

Internal pump modifications.

pump

washer

washer

incorrect

correct

Correct and incorrect washer to shaft location.

and, by setting it up in a vice, I was able to press the shaft back into the rotor a distance of 225 thou, which was the amount by which the cam originally protruded to excess from the front of the block. Then 225 thou were ground off the non-splined or tail-end of the shaft, enabling it to seat properly within the pump body. Much to my relief, on assembly, I now found that everything was OK, and this proved to be the case when the engine was run.

I do not think that these modifications in any way reduce reliability, as the area over which the washer and shaft splines make contact is more than adequate to prevent shearing and is almost certainly still stronger than the earlier pin/slotted shaft arrangement. The reduction in bearing area for the shaft, resulting from the shortening of the pump boss, is such as to be almost negligible bearing in mind that it is swamped with oil. The very nature of the spline drive imparts a sort of universal joint property and prevents, or at least greatly reduces, any side-thrust loadings on the shaft or bearing surface of the boss.

No doubt, a couple of my earlier remarks have set some of you thinking, namely that I was using one of the latest camshafts in my 850 racer, and that I considered the pump had a serious design fault.

On the first count, I have since reverted back to a 649 cam, for although the later cam, C-AEG597, gave quite good results, it was no better than the 649 on this particular engine and, in addition, needed some 14° static ignition advance on a Cooper S distributor. And I feel that to run an engine on this much advance spells trouble. Besides which it ruins top-end performance. 10° static advance on this distributor is more than enough. On the second count, I don't think that this will cause trouble on 850 racers which only develop between 85 and 90psi oil pressure, nor will it cause trouble on those engines using thin mineral oils. However, in my opinion, the fact that the pump shaft is only an unkeyed press-fit within the pump rotor could easily cause trouble on those engines using vegetable castor-based oils (the ideal oil for a racing A series engines) which are very thick and heavy when cold. At such times, I feel that the resultant forces applied to the pump could well cause the shaft to rotate within the rotor without actually turning it. On the other hand, it should not prove too difficult to drill and pin, or make up a keyway to prevent this twisting of the shaft within the rotor. Certainly, I shall endeavour to do this when I find myself using these latest components.

I should point out that I have recently been told that there is a pump available with spline drive which does not necessitate all these fiddly

56

The Minis tail the Fords round Brands Hatch.

mods when fitted to an 850 or other similar A series engine. Frankly I should say that to date I have been able to find no trace of such pumps in BLMC part books and neither has my local distributor. So this can only be treated as hearsay. Further, the automatic Minis utilise yet another type of pump, which won't fit other Mini variants.

One small final point on fitting the splined pump to my particular 850. The pump body is somewhat larger as far as external dimension is concerned, and I found that it fouled against the bell-housing. A couple of minutes with a file soon cured this, however; some later bell housings have small cut-aways to give sufficient clearance, but the difference is not such as to warrant expenditure on a new housing.

chapter five

Gearboxes

Some people are under the impression that one can fit close ratio gears to almost any Mini for as little as £15 0s 0d. This is not strictly true. If the box is of the baulk-ring type then one can fit the 997 Cooper gears which are slightly closer ratio, or if of the needle roller type then standard 998 or Cooper S gears can be used both of which have slightly closer ratios than ordinary Minis. (Workshop Manual gives full details.) It is my opinion that such gears unless easily and cheaply available, are rather a waste of time, certainly at £15 0s 0d they are.

For £25 10s 0d one can obtain from BLMC Special Tuning Dept. a set of ultra-close ratio gears either straight cut or helical—the former are the best bet providing one can tolerate the extra noise. These gears can be fitted to almost any of the transverse A series boxes, though in the case of non-Cooper S or pre 1966 units it will be necessary to carry out certain machining operations and to purchase several extra parts to replace existing parts before the latest gears can be fitted. Just what needs to be done will depend upon the exact vintage of the existing box, those prior to 1963 needing more work than those that followed. Again a study of the official BLMC parts book will show just what is needed.

Whilst on the subject of gearboxes and their interchangeability I think it is worth pointing out that although there is little difference between the various transverse boxes providing they are of the same vintage, there was up till Sept 1967 one big difference between those fitted to the Cooper S and those fitted to other transverse A series units, even though since 1966 all had been of the needle-roller variety. To cope with the extra stroke thicker crankshaft different journal alignment the Cooper S box was made somewhat wider to give sufficient clearance to the crank. Many people who tried to fit S engines to ordinary Mini boxes soon discovered that the S crank fouled the sides of the box casing. If trying to fit the 1071 or 970cc units then in most cases all that was necessary was to grind a little metal from the inside of the casing. Problems arose however when trying to fit the 1275 unit for here the amount of metal needing to be removed was excessive. One or two people, my pal George Lawrence for one, had their boxes widened by cutting them and then Argon-arc welding back together so as to widen the casing. This posed many problems not the least being distortion of the casing. Fortunately it was not long before the manufacturers started to make all gearboxes alike in respect of major dimensions.

Whilst on the subject of close ratio gears let us take the opportunity of studying briefly the effects of fitting them. As you know when one changes gear the revs for a given speed either rise or fall depending on whether one changes up or down. Further, if one changes say from a high speed in top into an excessively low gear then the engine will blow up. Conversely if one changes from a low gear into an excessively high gear the engine revs will fall right off and the engine will tend to stall. Obviously this is factor of the differences in ratio and engine rpm for a given speed in either a high or a low gear. The smaller the difference between two gear ratios then the smaller will be the variation in engine rpm for a given speed when changing from one gear to another.

Standard production engines will work over a relatively large rev. range of about 5000rpm. Competition racing engines usually have a much smaller range of about 1500rpm. Thus on standard cars one can tolerate large differences between adjacent gear ratios because a drop of say 3000rpm does not matter, but on a racer one cannot

Straight cut, five speed Mini gearbox.

B.M.C. works group two rally Cooper S.

tolerate this. Unfortunately for us Mr Average likes to pull away from rest at, say, 800rpm and yet reach 80mph on the motorway. The former demands a very low first gear and the latter a relatively high top gear. Thus four forward speeds are spread over a large ratio range on standard cars, which do not feel the resultant effects of big differences in rpm from gear to gear for a given speed. But this is no good for our racer. A drop of more than 1500rpm makes it tend to stall. Since top gear nearly always remains constant—gearbox ratio, not final drive ratio—we must resort to altering the ratios below this. So one makes bottom a bit lower than the standard second and crams the remaining 2 gears into a ratio gap previously only occupied by one gear, thus reducing the ratio gap between adjacent gears and hence reducing the rpm drop from say 3000 to 1000. Simple, isn't it? On big engines with bags of torque the answer is usually yes, but on small engines, no. Bottom is now too high to allow the vehicle to make a reasonably easy and quick get away from rest. The answer? Put in a fifth gear somewhat higher than the standard first gear. The other alternative to this: lower the final drive gearing (raise the ratio) thus lowering the overall gearing. Marvellous but now at max rpm we are 20mph down. The answer: Put in an overdrive top.

Basically it boils down to this. The narrower a car's usable rev range, the more gears and the closer the ratios you need in your box, and one usually needs a fairly low bottom gear.

NB. There is little point in fitting ultra-close ratio gears to a standard engine since this usually makes it overgeared in bottom and difficult to pull away in, or greatly undergeared in top, reducing mph and mpg. But not all racers with small engines have 5 speeds. For cost reasons many small-engined saloon racers (eg 850 Minis) still use 4-speed boxes coupled to a very narrow rev range. One simply has to tolerate a lot of clutch slip and a most undignified departure from rest under such circumstances, unless one is twit enough to undergear oneself in top.

Whilst on the subject of gearboxes, I have been asked by several

John Buncombe leads the pack into Devil's Elbow at Lydden.

people about fitting the Mini Cooper remote gear change to the ordinary Mini. This is quite OK but it is not the simple task often imagined. It does involve removing the gearbox and fitting a new casting to the back of the final drive housing of which, when bolted up, it forms an integral part. Theoretically this involves certain line boring operations, but in practice many enthusiasts have found this unnecessary.

I have deliberately avoided mentioning the latest all synchromesh gearboxes as fitted to the latest Cooper S, the 1300cc versions of the BLMC 1100, and the very latest long-nosed version of the Mini, the Clubman. I have done so for the good reason that although the boxes may look just like any other Mini box externally, being identical, in respect of change linkages etc, etc, they are internally very much different and must be regarded as complete departures from the earlier boxes. The only practical interchangeability between the two types lies in the differential and its housing provided that one bears in mind my earlier remarks about the accurate mating of the diff-housing to the main gear casing.

The close ratio gears already mentioned will not fit these all-synchro boxes and it is not a practical proposition to modify the box to make them fit, neither can the non-synchro gears or other major components be introduced into an all synchro box or casing. For a while this meant that if one purchased the late Cooper S for competition one was forced to throw away the existing box and fit an earlier unit complete. Fortunately BLMC Special Tuning have recently put close ratio gears to suit the latest boxes on the already comprehensive list of tuning accessories available from Abingdon.

One final point worth noting is that since production of the Mk 2 Mini and now the Clubman type has really got under way all boxes are merely modifications of those boxes designed to accept a remote control change. That is to say that even those Minis still produced without remote changes, including vans, do in fact utilise the remote change type box casing; the remote change mechanism is absent and in its place one finds an adaptor mechanism enabling the earlier 'magic wand' change to be fitted in its place. To convert to remote simply involves removing the adaptors, cutting another hole in the floor and bolting the remote mechanism into place.

However, despite this latest innovation allowing even greater interchangeability my remarks about regarding the all synchro unit as a separate box type altogether still apply; their internals will not fit even the latest three speed synchro boxes, the latter remaining as earlier units internally. It is however true that the all synchro box will mate up to *any* of the transverse A series range of engines. Frankly I have never considered this really worthwhile particularly if competition is intended. The extra complication is just something else to go wrong and in my opinion no self-respecting racer or rally driver would rely on synchromesh to help change gear anyway.

Some people still think that the 1100 and Cooper remotes are similar and are interchangeable. There is however a major difference. The Cooper, Cooper S and now Mk 2 Mini remotes bolt direct to the rear of the differential housing using four bolts which screw into the threaded diff. housing. The 1100 set-up however utilises a thick rubber block which fits between the forward end of the remote tunnel and the back-end of the diff. housing Coupled to this the two components fit together at a different angle from that on Mini versions. Unfortunately, for competition use the rubber block arrangement renders the change rather less than ideal introducing a considerable loss of precision to the change. I well remember a friend of mine who races a Mini-7 Formula car complaining that he could not select gear properly since he had fitted an 1100 gearbox to his car, these usually being a little cheaper on the second-hand market than the Cooper equivalent. I suggested that he might replace the standard rubber block with a steel or alloy block thus making the joint rigid. This proved entirely successful and completely cured the selection problems. Since that time I have tried it on several other cars which were experiencing selection bothers with 1100 boxes. In all cases the remedy has proven itself 100%.

Final Drive Ratios

Following on from my exposition on the use of close ratio gears I would like to consider the application of different final drive ratios. First remember that the higher the final drive ratio, the higher the vehicle's overall gearing and vice versa. For example a Mini fitted with a 3.765 to 1 final drive is more highly geared than one that is fitted with a 4.133 to 1 ratio.

It seems a common misconception that one only has to fit a higher final drive ratio to improve the overall performance of a car. Often quite the reverse occurs, because the vehicle becomes overgeared.

This becomes easier to understand if we consider our childhood days of bicycle travel. I well remember my bicycle which had three gears. To ride up a hill I needed to use bottom gear because I lacked the strength to push the pedals round in top gear and going down hill I could not pedal fast enough in bottom gear and hence had to use top gear. Similarly we have cars with four forward gears. The overall effect of these gears is governed by the final drive ratio. If this ratio is too high then the effect is exactly the same as trying to get up hill in top gear, which effectively prevents a vehicle revving at its optimum speed in top gear, or unduly slows down the rate at which revs can increase in the indirect gears (reduces the rate of acceleration). Conversely if the final drive ratio is too low the vehicle becomes overgeared and cannot rev quickly enough (without blowing up) to give the vehicle a reasonable top speed.

Normally speaking standard cars are fitted with the ideal final drive ratio; any alteration can only spoil the overall performance, unless special factors exist which justify the fitting of different ratios. For example some people may regularly use their car for towing heavy loads; under such circumstances a lower final drive may be justified. Similarly the demands of motor racing or rallying must dictate an axle change to alter a vehicle's overall gearing.

Remember this however, if we lower the overall gearing, although we may improve acceleration times up to maximum speed, this maximum speed will be less, unless the engine can and will rev more highly. Thus for example, many racing Minis have only got top speeds of 105 to 110mph and need to rev at 8000rpm even then. Acceleration to this speed is, however, fantastic, due to the use of a low final drive ratio. Normally one strikes a balance between top speed, acceleration, and permissible and attainable engine rpm.

If one increases a standard vehicle's engine power, conditions may demand the use of a higher final drive ratio. Such conditions usually are continuous flat out driving in top gear and/or the desire to reduce engine rpm for a given speed, which often improves petrol consumption. One must always sacrifice some acceleration to attain these objectives. Personally, for road use I would keep a standard final drive ratio unless at least 60 per cent of my motoring was on motorways and autobahns, *even if* I have a tuned engine, and *definitely* if I have an untuned engine.

Perhaps I should add one small qualification to this statement. Cars with engines over $1\frac{1}{2}$ to 2 litres are often capable of pulling a higher gear than that which is fitted as standard. Hence the use of overdrives, which allow a car to use its manual forward gears up to maximum required speed. Then the flick of a switch effectively lowers the back axle ratio raising the overall gearing, resulting either in lower engine rpm for a given speed or higher speed for a given engine rpm.

Quite often these overdrives operate not only on top gear but also on the intermediate gears. Thus a four speed box effectively becomes say eight speeds.

Interchangeability of final drives between the various transverse A series units is complete. That which fits the first Mini ever made will also fit the very latest 1300 variants or the Cooper S, and vice-versa, the only qualification to this being that as the crown-wheel and pinion are a matched pair, one must use the correct wheel and pinion pair. Ratios 3.44 to 1 up to 4.1 to 1 (standard 1100) have different pinions for different wheels, but from the 4.1 ratio inclusive down to the lowest gearing listed, all the different wheels utilise the same pinion, namely 22G99.

If one includes in these considerations those crown wheels made especially for competition then the picture is a little different. Up till mid 1969 the limited slip differentials used in most racers were designed for use with any of the standard or 'standard equivalent' crown wheels. However since that time many people have changed to a Salisbury type LSD unit and here lays the problem. The Salisbury unit cannot be used in conjunction with the standard crown wheel type and although some people merely altered the existing wheels quite the best solution is to obtain from the Special Tuning Dept. a diff suitable for use with the Salisbury unit. Pinions remain unaltered.

Bowden Racing's glassfibre fronted Mini.

Tony Lanfranchi chases a wildly understeering John Rhodes.

Britax-Cooper-Downton, Morris Cooper S.

Two early Mini racers rounding Druids.

chapter six

Interchangeability

It has often been said that the Mini has reached the end of the road as a successful racing car, not only internationally but also at club level. Anglias and Imps are generally regarded as the substitute for Minis. How is this? Apart from works cars participating at club levels 1300 Cooper S still reigns supreme. It is true that 1000cc Imps and Anglias can beat 1000cc Minis, but only when works-prepared. In the 850 class, an Imp will at times emerge victorious, with a lap time at which present Minis cannot approach, but not very often.

Nonetheless, for all the pessimistic talk and defeatist statements that the Mini is 'over the top' or 'obsolescent, old chap', Minis still win races and plenty of them, in all three classes, albeit mainly in Club racing. Also, more and more newcomers are making their debuts in Minis and old stagers are building more powerful and lighter-in-weight versions. None of this points to the Mini being a car which is past it; yet we all know that Imps and Anglias are in fact quicker. (I will ignore the 1300 class for the moment). Why then is it that Minis still win races and command the attention of the Club racing enthusiasts? Well, the answer to the first question lies partly in supremacy of numbers and partly in care of preparation; the answer to the second question is easy, cost.

Though Imps and Anglias can be made to go very, very quickly indeed they have one big drawback. They are fantastically expensive to prepare. I would almost guarantee that there is little change out of

£2500 for a Fraser Imp and little change out of £3000 for a Broadspeed Anglia. Certainly figures well beyond most club enthusiasts. Unfortunately to spend much less than this on either of these cars usually results in an unreliable, half-sorted, half-competitive racer.

On the other hand Minis are far cheaper to prepare. A fully-prepared 1000 or 850cc Mini only costs between £800 and £1200, sometimes far less, depending on how you purchase your bits and pieces and whether you are preparing to Free Formula or Formula Mini 7 regulations. Some quite reasonably competitive Minis have been prepared for as little as £450. Obviously this is the car for the average clubman. Still this doesn't explain why this is true. Why are Minis so much cheaper to prepare? Are BMC spares much cheaper than Rootes or Ford? Certainly not. I wish they were. No, the answer lies in the words interchangeability, availability and flexibility.

Let us consider them in reverse order. Flexibility, what do I mean by this? Simply the number of standard production parts, manufactured by the makers themselves and on ready sale to the whole public as original equipment or optional extras, which can, safely and successfully, be used in standard or slightly-modified form for racing. For example, optimum results on an 850 Mini are obtained using an 'ordinarily modified' standard BL cylinder head, exactly as fitted to several of their A series models and obtainable over the counter from any distributor. On all Minis standard production crankshafts are quite suitable, though admittedly it pays to have them toughened or hardened if they are not S or ex-Special Tuning Department.

It is possible for any reasonably competent enthusiast to modify a dry Mini suspension to full race specification for 50s plus the cost of a set of shock absorbers, which for a Mini are cheaper than for other cars anyway. Compare this to the Anglia's downdraught cylinder heads (must cost at least £150), steel crankshafts and rods (Minis don't need steel rods) and one-off suspensions, with wishbone front ends and coil spring rear ends. All very special, all very much one-off and all very, very expensive. Even the Fraser Imps have separate types of heads for club racing and international racing, neither of which, I gather, are readily available as a standard fitting or optional extra. I admit that full-race Imps *seem* to be a little more standard than full-race Anglias, but there isn't much in it. No, one of the beauties with a Mini is that 95 per cent of the parts one uses are standard BLMC production items, sold at standard production prices, not inflated special one-off prices.

Although availability has been partially covered under flexibility, in that as I have said 95 per cent of one's spares are readily obtainable over the counter, this does not take into account the second-hand or scrapyard market. There are so many Minis on the British roads today, and have been for the past five or six years, that due to sheer weight of numbers, thousands have been involved in accidents, written-off by insurance companies and dumped at scrapyards. Sometimes one hears of a write-off Mini in time to put in a bid to the insurance company. If lucky one may finish up with a bodily badly-damaged Mini mechanically not too bad and two-three years old for about £20-£50. An extremely cheap, useful and comprehensive set of spares. On the other hand this is the exception rather than the rule and more usually such spares are purchased from the scrapyard where one usually has more than one vehicle to choose from and a chance to knock the 'scrap man' down on his asking price. Examples of how cheaply one can buy spares at a scrapyard are:

Article	Scrap price	New price
SPQR steering col adjuster	1s	12s 6d
2 Konis for rear of Mini (almost new)	30s	£9 plus
Mini radiator	30s	£7

There are so many Minis about that no matter what you need, you can almost guarantee to find it at one of the local scrapyards. The fact that more people seem to tune and modify Minis than Imps and

Anglias is a great help when after special shockers, etc. In any case not much stuff found on Imps and Anglias is any good for racing.

Finally on this question of availability. The fact that so many people modify, race, rally, autocross, hill-climb or sprint in a Mini means that there is a flourishing second-hand market for the more popular and specialised spares, outside of the scrapyards. Things like Cooper S valves, S drive shafts and discs. 1100 cylinder heads, straight-cut gears are regular items to be found in the miscellaneous columns of any motor magazine with a sporting bias.

Then on top of all this is the new one-off market. Almost every town and village has its backyard manufacturer of some Mini bits or pieces. Some of their products are good, some not so good, some downright dangerous, but one thing is certain, they are prolific. This, if nothing else, means that prices must be competitive. In 1961 you could buy a rough glassfibre Mini bonnet and boot panel for a little over £10. Now, however, these panels are extremely well finished and sell for between £3 or £5 per pair by several vendors. Compare these prices with those for similar articles on an Imp or Anglia.

Now we must consider how the interchangeability factor (I) exerts an influence and take a closer look at some of the more technical aspects not previously covered in these pages. We decided that flexibility and availability (F and A) as related to the BLMC Mini and its derivatives meant a readily obtainable supply of spares suitable for use in racing. Since most of these spares and goodies are off the shelf BLMC spares, not only are they reasonably readily available but are fairly economically-priced, or downright cheap if one goes to the flourishing second-hand market. These factors could not be so important were it not for 'I'. Why is this, and what does 'I' mean?

Interchangeability means that it is, for example, quite possible to take certain parts from the race specification Cooper S and *without modification* use them to replace a somewhat weaker component on, say, the 850 Mini (eg, all reciprocating valve gear), thereby increasing the reliability of this engine and consequently its potential usable power output.

How does this influence 'F' and 'A'? Well these two factors would be no good on their own other than on the vehicle for which they were designed. Since the Cooper S and, to a lesser extent, the 998 Cooper are the only models which can, using their own parts, be safely raced, one would be unable to race an 850 Mini reliably unless one could use a fair proportion of Cooper or S parts or expensive one-off equipment. Thus no matter how good the 'F' and 'A' factors are, one is still limited to say Cooper S racing unless one has the 'I' factor.

The same arguments apply right through the BLMC range and explain why, for example, one sees more Midgets and Sprites than Spitfires in Modsports racing. I have discussed BLMC engines and their interchangeability so many times in the past that I do not intend to bore you with the details again. Back reference to the Mini Tuning book should answer any questions in this direction. However, I have been asked to discuss the 'I' factor as it relates to body/chassis/suspension/brake units. The following attempts to throw some light on the matter.

Taking first the body. Basically *all* Mini bodies are identical as regards size and shape. The mounting points for all mechanical appendages are identical excepting that some models have the hole in the floor for the gear-change in a different place, but this can soon be changed using a pair of tin-snips and an aluminium blanking plate. The early Minis from 1959 to approximately May 1960, were some 40lb lighter due to thin gauge steel being used in the floor and seat, together with the absence of one or two minor structural plates which usually pass the inexperienced eye without being noticed. The rear shock absorber mounting points on 1959 Minis were rather weak and had a nasty habit of giving way, letting the shock absorber up into the boot. No safety belt mountings were included till about 1962. Since 1965 the front valence has been cut away each side to allow cold air to get to the brakes, an improvement brought about by the actions of clubmen. Even the Elf and Hornet bodies are identical to the Mini as regards major dimensions, and accept the same mechanical components including sub-frames and suspensions, etc.

There are no differences between Hydrolastic and solid rubber-suspended Mini body shells. If you write off a solid rubber Mini you rebuild it using the same shell as the chap with hydrolastics. One point to remember. Since the September 1967 announcement these remarks are qualified by the fact that the latest Minis have larger rear windows and one or two other minor alterations.

Immediately many of you readers will say 'The man's mad, what about the latest long nosed variants, the Clubman and the like?' Well, the answer to that is that in actual fact the differences are much less than they at first seem, though there are no door hinges on the exterior the shells are identical from all practical viewpoints from the front bulkhead seams backwards. So similar are the shells that the ordinarily shaped Minis can be converted to Clubman shape merely by fitting a new front end and in fact there are, on the market, a few glass fibre conversions which enable one to do just this at a very reasonable cost. All major mechanical parts including sub-frames, etc, can be regarded as identical.

However, for all practical purposes all Mini shells are identical. If one is using *new* sub-frames one uses the same frames front and rear whether the suspension is solid or hydrolastic. However, the earlier solid-suspension Mini frames were different in that there were no holes drilled in them to take the hydraulic hoses used on hydrolastic suspensions. The latest frames *all* have these holes. Otherwise *all* sub-frames can be treated as identical. The frames are the same whether the vehicle be van, traveller, Elf or Hornet, Mini, Cooper or

Cooper S. Suspension-wise things are a little more complicated and we need to delve a little deeper into the problem.

Taking first the moving parts of the suspension:

Rear Suspension

All major dimensions are the same and are interchangeable but latest equipment is slightly different and in some cases superior. The massive iron swinging arms, which carry the rear wheels, are identical throughout the *entire* Mini range excepting that about 1962 the bushes on which the arm pivots were replaced by needle rollers. Anyone building a Mini for competition these days would be well advised to use this type as the wear rate on the bearing and the steel shaft which runs through these bearings and which supports the arm, is far less than with the early system of bushes. In fact the early system of bushed type of arm is no longer manufactured.

The suspension struts or trumpets were originally fabricated from steel, but were soon replaced by cast alloy, which is not necessarily superior. Again these struts are completely interchangeable on solid suspensions, though those fitted to vans and travellers are 200 thou longer and need reducing when fitted to saloons unless ride height is to be excessive, but five minutes with a hack-saw, on the narrow end soon cures this problem.

Hydrolastic Minis are fitted with rods instead of trumpets. These act in exactly the same manner and fit in the same way, but are not interchangeable with trumpets.

On some early Minis small round wire washers are found between the ends of the trumpets and the shoulder of the hardened steel ball which fits into the end of the trumpet. These are dispensable, and were merely fitted to increase the ride height. When lowering a Mini these can be thrown away, or if one wishes to raise the suspension one extra washer may be fitted. The rubber suspension cones on all solid Minis etc, are theoretically identical, though very early 1959/60 cars had a different rubber mix, giving better road holding characteristics: it is impossible to recognise these cones, the part number is unchanged, and one can only hope that if one has a 1959 Mini it is fitted with the original rubbers.

On Hydrolastic Minis the rubber cone is replaced by the hydrolastic unit, and they are interchangeable providing one uses Hydrolastic rods instead of trumpets and all the other hydraulic paraphernalia which comprises the Hydrolastic suspension. Apart from the rubber cones of Hydrolastic units no suspension parts are interchangeable front to rear.

Front suspension
Unfortunately there is not quite such a marked degree of interchangeability on the front end of a Mini when compared to the rear, where pretty well all parts are interchangeable from model to model. However, things are not too bad and there is quite a fair bit of scope for imagination.

All rubber cones are interchangeable on the 'solid' Minis and variants, as also are the front struts or trumpets. In other words the remarks made for rear suspension, including those made on hydrolastic suspension, apply in exactly the same way at the front end. The top suspension arms, including those on Hydrolastic Minis, are all interchangeable and similar, excepting that the latter do not have any shock absorber mountings on them. This is easy to rectify, the arm itself being similar.

The bottom suspension arms are again, without exception, all identical, Mini, Cooper S, Hydrolastic, the lot. The only thing is that the latest models, particularly Cooper S, have larger eyes at the inner end so that they can accept the latest rubber bushes, which have steel inserts to prevent them breaking up and, ostensibly, to give better location. Otherwise these arms are the same. If one has the latest arms with larger eyes, one merely uses the latest steel inserted bushes; on the other hand, if you have the smaller-eyed type then the easiest

way is to use the earlier all-rubber bushes, though it is possible to use
the steel insert type if you persuade these bushes into the smaller eye
using soap solution and brute force, eg. a vice. It is these arms that
are lengthened to obtain negative front wheel camber. The tie bars,
that is those bars which join the front of the sub-frame to the small
end of the bottom suspension arm, are all identical on all Minis. By
varying the thickness of the inner rubber bush, and/or washers
between the car and sub-frame, one can alter the front wheel castor
angle.

The steering arms, that is those cast, cranked arms which link the
hubs to the ends of the steering rack, are again interchangeable but
there is a difference. After the first few Cooper Ss were produced it
was felt that they needed beefing up and on the Cooper S only, the
diameter of these arms was increased, which gives food for thought
if one is racing a Mini on ordinary steering arms.

I should perhaps mention in passing that on Hydrolastic Minis it is
possible to obtain special competition units from BLMC. Most people
seem, however, to prefer to use rubber suspension, which is easier
to set up and is some 100lb lighter. On the other hand, probably
the quickest Minis ever produced, the Alec Poole Hornet, the
Janspeed SCA and the Harry Ratcliffe Vita-D 1300 fuel injected club
Minis used Hydrolastic suspension, though Alec told me at
Silverstone that he was thinking of changing back to solid suspension.
Steering racks are all the same.

Brakes
In this field one has a very, very wide choice if one takes into account
the varying modifications to wheel cylinders, brake linings, etc.
However, for the sake of simplicity, for all practical purposes one can
make several broad divisions based on brake size and model of car.

Narrow drum brakes:
These were fitted to all Mini saloons, vans and travellers up to 1964,

and were of the single leading shoe variety. If using an early 1959 or 1960 Mini it pays to fit the 1964 wheel cylinders used on the last of the single leading shoe type. They can be fitted as straightforward replacements.

Wider drum brakes:
These were fitted on all non-Cooper Minis, vans and Travellers after September 1964. Elfs and Hornets have always used these brakes, though I was told that some of the very early ones were fitted with the narrower type. These wide drums were of the twin leading shoe variety on the front. It is quite possible to fit these wider brakes to those Minis normally fitted with narrower brakes, but one needs to fit them complete with back-plates and wheel cylinders, and the extra brake pipe. They cannot be fitted to the earlier back plates.
In actual fact the drums on all drum-braked Minis are identical. It is only the brake shoes which are wider, they just simply rub on a greater width of the drums. However, if carrying out a conversion one must fit new drums as well, since the narrower shoes will have grooved the drums, that is unless one is prepared to skim a fair bit (up to 30 thou) off the drum, and I proved that this is unsafe, at least for racing. Frankly if doing this I would fit Minifins to the front, and ordinary drums to the rest.

Cooper Mini:
These are fitted with similar brakes at the rear to other drum-braked Minis. Front brakes, however, are disc. These can be fitted, complete with the hubs to any other non-Cooper S model, without needing other alterations, apart from different brake fluid and possibly a larger master cylinder. Although neither of these disc set-ups are any better than drums, the brakes used in the 997 Cooper were worse than useless. Those fitted to the 998 Cooper are much better, having larger calipers which can be easily fitted to 997 Coopers. Nonetheless I prefer to regard ordinary Cooper brakes as being of no practical use as a conversion for drum brakes and completely inadequate for any 1000cc or greater, racing Mini.

Cooper S:

Again a disc/drum set-up but much better, in fact quite good. Rear drums incorporate iron wheel spacers and the wider brake linings are used. However, these drums are very heavy and I would prefer to use ordinary Mini drums or Minifins with alloy spacers, which would show a considerable saving in weight.

The Cooper S disc front brakes cannot be fitted to any other Mini unless one also fits Cooper S drive shafts, hubs and master cylinder, etc. Even the flexible brake hose is different, being stronger and less likely to swell under pressure, this being a constant source of brake loss on early S's. The latest type have a double green band for recognition, as opposed to a single green band.

Hubs:

There is little to note about the wheel hubs and bearings, except that the number of splines on early Mini drive shafts differed from current Minis, and consequently front hubs were different. Also, the Cooper S has Timken-bearings front and rear, though until September 1967 plain bearings were still fitted to the rear. Also the first few hundred Cooper Ss were fitted with an unsatisfactory hub/taper roller arrangement on the front, and were subject to excessive wear. This was soon modified however, and these bearings are now a great improvement over the plain bearings fitted to other Minis. Needless to say, the different bearings utilise different hubs. Disc brake set-ups use different hubs.

Drive shafts:

Other than the aforementioned differences in the number of splines there is little of interest in Mini drive shafts. There have been no radical improvements or alterations since the Mini's initial production, though the rubber cross drive shaft universal joints have been replaced by Hardy-Spicer joints on the automatic versions. It is not possible to fit these drive shafts to manual Minis, neither is there any practical advantage on anything other than the much larger

Cooper S, and these are now fitted as standard to the 1275 version. They are not, however, the same as on the automatic Minis, and are not interchangeable.

If one is fitting a limited-slip-diff to a Mini then no matter what type it is one must utilise the Hardy-Spicer set-up as used on the Cooper S, unless one is using the very early type of diff unit designed for use on the rubber UJ type set-ups. Unfortunately due to the necessity for changing the driving flanges from the diff unit as well, one finds that it is also necessary to change nearly all the pinions and shafts, etc, within the diff unit itself thus incurring much extra expense. Therefore unless using an LSD I would avoid using the Hardy-Spicer set up if I could. On the other hand if one is thinking in terms of racing then although I think that the rubber cross type UJ's are quite adequate for 850 Minis I do feel that the more rigid Hardy-Spicer units are an advantage on 1000cc racers and of course those of larger capacity.

If one wants to fit a Cooper S front end to a vehicle equipped with rubber UJ's then it does not matter that the S shafts are fitted with the Hardy-Spicer couplings. These can easily be slid off the shaft and the rubber cross type flanges fitted in their place. Similarly rubber cross flanges can be replaced by Hardy-Spicer couplings provided the differential is suitably modified to accept them.

Though it is quite easy to fit Cooper S disc brakes to other Minis provided one uses the appropriate hubs, drive shafts etc, one will however need to increase the rear track to match that given by the S front end. The cheapest and lightest method is to fit spacers onto the standard drums, these spacers to be at least $\frac{3}{4}$in thick. You could use S type rear drums which have an iron spacer built in, but these are heavy and expensive and I can see no advantage in their use. In either case, it is necessary to fit longer wheel studs to the rear hubs. If the spacers are not more than $\frac{13}{16}$in thick, standard Cooper S studs are OK, but anything over this, and one must use special extra long studs as sold by most accessory shops.

Unless your car is post '67 Motor Show, do not use the latest S steering arms, as they are designed for the increased lock on later Minis. Use the earlier S arms if using the car for competition. They are stronger than Mini arms, though at a pinch one could use the early 850 Mini arm if road use is all that is contemplated (these are after all the same as the very first 1071 S arms).

Due to the configuration of the S layout the steering arms are positioned further away from the longitudinal centre line of the car than were the originals. Consequently the car needs re-tracking, and here's the rub. The track-rod ends may need to be screwed so far off the rack tie-rods that not enough thread remains for the two to remain safely together. This is more true of those cars that have front wheel negative camber, including the standard Cooper S (the steering racks all have the same dimensions). At least $\frac{3}{8}$in of thread must remain to hold track-rod and tie-rod together. Failing this, there are two possible solutions. Obtain two new track-rod ends anyway—the originals will probably have corroded—and two new locking nuts to suit, screw the locking nut onto a suitable bolt and screw the track-rod end on after it till the two come tight together. Then get them nickel-bronzed together making them into one unit, longer than the original track-rod by the thickness of a locking nut. Fit and lock onto the rack tie-rod in the normal manner. This method has been proven quite safe and reliable by many racing enthusiasts. However, if serious rallying or autocross/rallycross is envisaged, I would recommend this alternative but more expensive method. Remove the tie-rods from the rack and get some longer special rods made to take their place.

Well that just about wraps up the 'boring' side of racing car preparation as it relates to Minis, though I should make one amendment to what I wrote earlier. I said that *all* Mini steering racks are the same. This is not quite true. Latest Cooper S racks are slightly modified, to reduce bush and kick-pad wear. A minor point, really, but worth remembering if one has to purchase a new rack.

Engine Interchangeability

The permutations are so great that I cannot cover them all but the following points are important.

850 Minis use connecting rods fitted with pinch-bolt small ends. It is quite straight-forward to fit the 1100 or 998 fully-floating bushed con-rods, but the pistons need to have circlip grooves machined in them, and 1100 gudgeon pins—or 948 Triumph Herald pins—are essential.

With the exception of the Cooper S and 1275 Sprite, 1300 etc, all con-rods are identical as regards major dimensions and if pistons are suitably modified on 850 Minis, are all interchangeable. Also the distance between piston gudgeon-pin bosses on 850 Minis needs increasing to take the wider small-end of the non-pinch-bolt rods.

Cooper S rods should not be used, other than on Cooper S engines. Further, there are two different types of Cooper S rod. The 1275 and 1071 are similar, but the 970 rod is longer. It's worth mentioning at this stage that the 1275 Cooper S blocks are taller than those in the 970 and 1071, which are identical. Also, latest Cooper S 1275 have strengthened main bearing housings and crankshafts are now cross drilled. The only Cooper S engine parts that are easily fitted to other A series engines are valve gear components and the cylinder head, though whether the latter is worthwhile is a matter for debate. It is true, however, that the Mk 3 Sprite 1098 block, which has 2in main bearing journals, opens the door to constructing hybrid engines based on major Cooper S components, but there is little gain on anything which requires a capacity other than between 1100 and 1200cc.

It is possible to fit 998, 997 and 1098cc engine crankshafts to the 850 Mini, but the main-bearing housings need thinning down to make room. Personally I think that this is a waste of time; it is cheaper and easier to change the whole engine as 998 and 1098 cranks (not 2in main type) are easily swapped, and this may be a useful exercise. However, pistons will need changing, 998 engines having shorter pistons due to the longer stroke.

Remember the following:—
(1) Other than Cooper S, all BLMC A series blocks are the same height and similar in all other outside dimensions and most internal ones.
(2) Capacity is varied by alterations to bore and stroke on 850 engines, and stroke only on 998 and 1098 engines. If the stroke is reduced, the piston is taller and vice versa.
(3) Compression ratios are varied by varying the dish in the top of the piston.

To simplify this problem let us endeavour to construct some purely hypothetical cylinder assemblies.

(1) 850 block, crank, 1100 rods and pistons. Rods, block and crank pose no problems; most blocks can easily be bored to take 1100 pistons. So far so good, but—these pistons were originally designed for a stroke of some 83mm. The 850 stroke is 68 odd. Thus we have a stroke difference of approx 15mm. Since blocks have the same height and rods are identical (apart from pinch bolt etc) the 1100

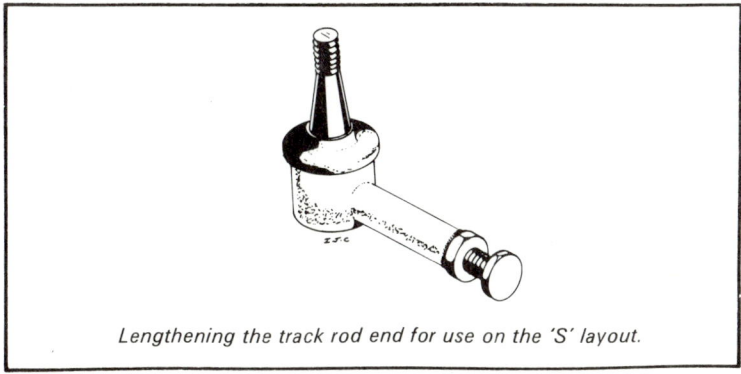

Lengthening the track rod end for use on the 'S' layout.

pistons must be about 7½mm shorter than 850 pistons, measured from gudgeon pin centre to top of crown, and obviously at tdc would not reach the top of the block, even if one shaved the maximum 3 or 4mm from the block face. Thus this engine cannot be built unless one has special non-BLMC pistons made.

(2) Sprite Mk III block, (bores as per 1100 and 998 engine) 1071 S crankshaft, 998 dished pistons, 1100 rods. As the Mk III block has 2in main bearings, and bore centres are not too dissimilar, the S crank and block will mate up ok. The 1071 S stroke was the same as an 850 Mini, 68mm or so. The Mk III Sprite has the same bore and stroke as ordinary 1100's, but the 998 is only 76mm and hence pistons are only 8mm shorter than 850 pistons, and to make such pistons reach the top of the block on this engine, one would only need to shave 4mm from the block face, and on some blocks this is possible. 1100 rods are ok.

Solid skirt pistons
I often receive letters from readers of *Cars and Car Conversions*, stating that they are carrying out X, Y and Z mods to their engine in an effort to improve performance. Among the more ambitious suggestions one commonly finds the use of flat top solid skirt pistons. Many people seem to think that this automatically increases power. Well, I am sorry to disappoint you chaps, but under certain circumstances, quite the reverse can be the case.
Let us consider each of the above mentioned piston characteristics separately, as they are unrelated.

Split skirt pistons do not necessarily reduce power or cause more friction. In fact many of the more common solid skirt pistons are fitted with skirt rings. From a power point of view these are inferior to those split skirt pistons which are not fitted with skirt rings, because they create considerably more friction. I am assuming of course that oil consumption is not a problem with its excessive carbon fouling of the plugs and chambers. If this is a problem and

88

performance is of prime importance, then I would prefer to re-bore and use oversize pistons than merely a change to pistons with skirt rings.

I agree however that generally speaking plain, solid skirt pistons are superior to split skirt types, but unless the skirts are ovalised and clearanced, then the only real gains are in reliability at higher rpm. There is not so much danger of piston breakage. It boils down to this. Split skirts don't necessarily cause power losses, nor do solid skirts necessarily cause power increases. However, extra piston rings increase friction, this does cause power loss. If an engine uses oil; do not fit extra rings. Improve the fit of the rings and pistons by reboring and renewing. If you are told that extra oil control rings on the skirt are the only effective way to control oil consumption remember this—new engines do not normally burn oil, nor do they use pistons with skirt rings. By reboring and renewing you should return to *new* specification. If an engine still burns oil then it has been bored or assembled wrongly or somesuch. In which case march it back. Make sure however that you are not using oil elsewhere, such as through leakage or worn valve guides.

Increasing engine capacity
Another common query concerns the increasing of engine capacities

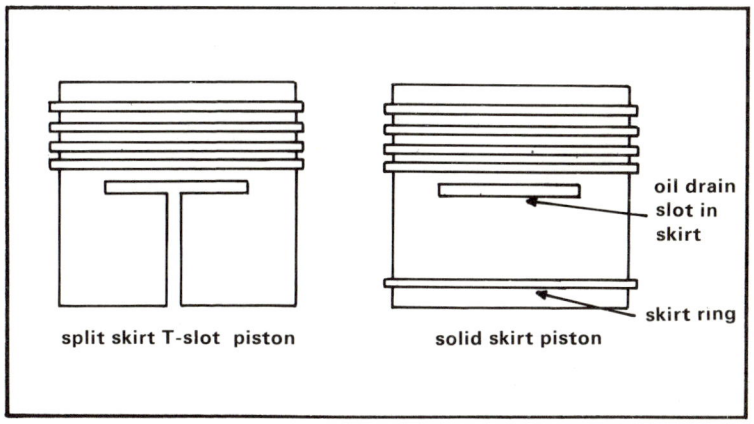

split skirt T-slot piston   solid skirt piston

oil drain slot in skirt

skirt ring

beyond the limits normally set out by the manufacturers. Normally this involves considerably overboring the block and/or increasing the crankshaft throw. The queries are usually either (i) Can I bore my engine to Xmm, (ii) If I fit such and such crankshaft do I need different rods and/or pistons, (iii) What is my resultant engine capacity, (iv) Will it be necessary to fit a different camshaft.

Taking these in order. A safe guide on the absolute maximum to which one can bore an engine is to find out just how far the conversion specialists bore the engines. For example, most people bore the Ford 105E type engine to 85mm. However, remember this— a fair percentage of engines bored to the supposed maximum are rendered useless by virtue of becoming porous somewhere in the cylinder wall, usually due to small casting irregularities at production. Thus no matter what engine, boring to the absolute maximum is at best, somewhat risky. When fitting a crankshaft with a different stroke from standard, providing the block is wide enough to accommodate the crankshaft the thing to remember is this. If the stroke is increased the

Con rods. Left to right; 850 Mini, Riley Elf and Cooper S.

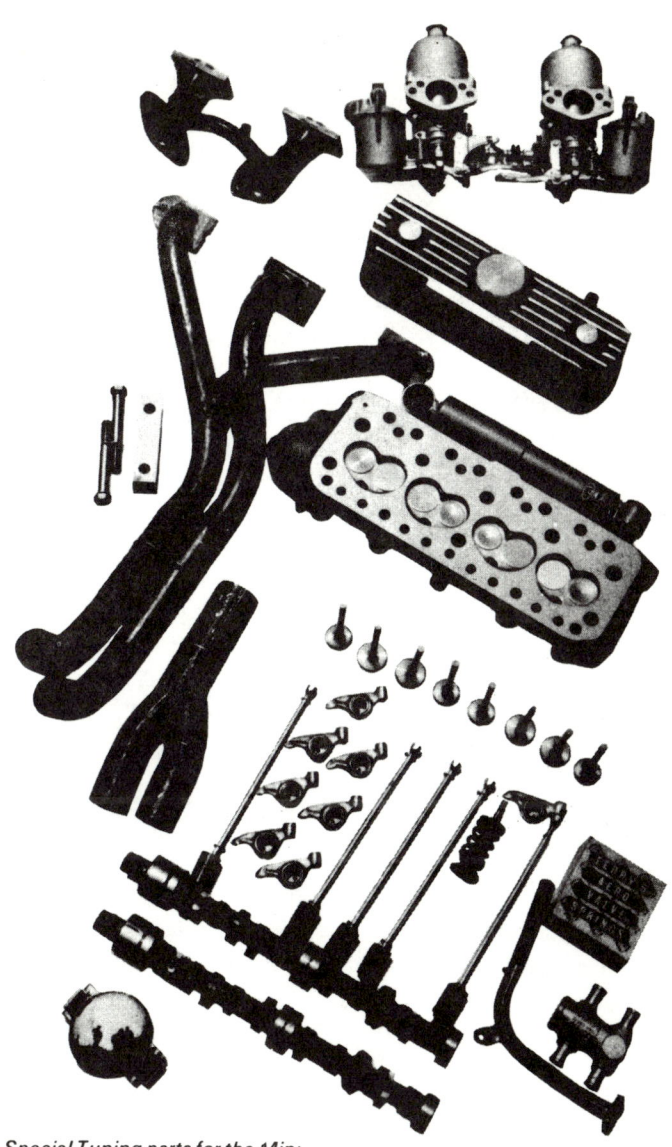

Special Tuning parts for the Mini.

pistons are pushed further up the bore and will tend to protrude beyond the top of the block. This must be overcome by fitting pistons with a smaller compression height (distance between gudgeon pin centre and piston crown) and/or shorter connecting rods. On the other hand, if the stroke is decreased then the pistons will not come to the top of the block and one must either shave the top off the cylinder block or fit taller pistons and/or longer rods. Not being a lover of mathematics, nor having unlimited time available, I cannot offer to calculate the various capacities arrived at by individual enthusiasts' modifications. However as with compression ratios this is an easy if somewhat laborious operation, one merely applies the following formula.

$$\text{Engine capacity} = \text{stroke} \times \left(\frac{\text{bore}}{2} \times \frac{\text{bore}}{2} \right) \times 3.14 \times \text{no. of cycls.},$$

$$\frac{\text{bore}}{2} \times \frac{\text{bore}}{2} = \left(\frac{\text{bore}}{2} \right)^2$$

Thus $\left(\frac{\text{bore}}{2} \right)^2$ is merely half the bore multiplied by half the bore.

$$\text{Number of cylinders in engine} = \left(\frac{\text{bore}}{2} \right)^2 \times \text{stroke} \times 3.14.$$

Thus the engine capacity is:

$$\text{Number of cylinders in engine} \times \left(\frac{(\text{bore})^2}{(2)} \right) \times \text{stroke} \times 3.14.$$

It is not necessary to use a different camshaft, but often advantageous, since generally speaking, large engines tend to give power over a fairly wide rpm range even when fitted with diabolical camshafts.

chapter seven

Ignition and generator

It is not the intention of this chapter to deal with material usually covered in articles on car electrics. I am assuming that you will be familiar with the practical and theoretical implications of the statements that plugs and points should be free from excessive deposits and set exactly to the maker's recommended gaps, worn or pitted points should be ground to give a flat sparking surface and point surfaces should mate up squarely with each other over the entire sparking surface.

If this is not so, and cannot be achieved, then new sets should be fitted. The entire electrical system should be clean and free from dirt and grease, though it is advisable to apply a little clean grease or water-repellent to joints and unions to keep water out and prevent ignition failure through wetting. This is a side of maintenance often neglected and I am guilty myself in this respect where my road car is concerned. It is nonetheless a topic well understood, even to the average motorist, and I can only say that on any highly tuned competition car, attention to this side of maintenance is essential if one hopes for any success. The following, therefore, deals exclusively with basic mods to existing set-ups.

Spark plugs
Enthusiastic drivers rarely use their cars in a manner conducive to long life of the maker's standard plug. Usually these plugs are very soft (hot) and are easily burned out. Invariably it is essential that at

least a slightly harder (colder) grade of plug is fitted, say N3 instead of N5, even if the plug type remains unaltered. Failure to do this can lead to rapid plug burning or even complete breakdown, causing misfiring. For more extreme circumstances it may, in addition to fitting a harder grade of plug, prove necessary to fit a different type— changing perhaps from extended-nose to standard gap or even, on racing engines, recessed-nose racing plugs. The more highly tuned an engine, the higher the revs, the higher the compression ratio, then the harder (or colder and tougher) its plugs need to be.

Generally speaking, recessed-nose plugs are tougher and have the most resistance to wear. Standard gap come second in this respect, while extended-nose have comparatively, least wear resistance. Combine the varying grades of hardness for each plug type and one can see that there is little difficulty in finding a plug exactly suited to one's requirements. It should be remembered that the harder or colder a plug the easier it will juice up at low engine speeds and the more difficult the engine will be to start from cold. Indeed, the mill itself may be damaged through continued cold starting. Thus for general use a compromise becomes necessary.

Coil
Many people think that immediately an engine is tuned it becomes necessary to fit a red-topped sports/racing coil. Maybe this is true on some engines but not on the BLMC A-series. In fact, not only is a standard Mini or Cooper coil adequate but it is essential if excessive points-burning or electronic rev counter consumption is to be avoided. It is true, however, that full race engines (eg with 649 cams, 13:1 cr) may benefit from the use of a heavy duty coil designed for six cylinder cars such as the Healey 3000. However, the red-topped out-and-out sports/racing coil should only be used on full race 1300 Cooper S units and even then it's a debatable point.

What is essential in a coil is that the HT lead should be connected to the coil through a screw-in cap, not just a push fit into the top. I have

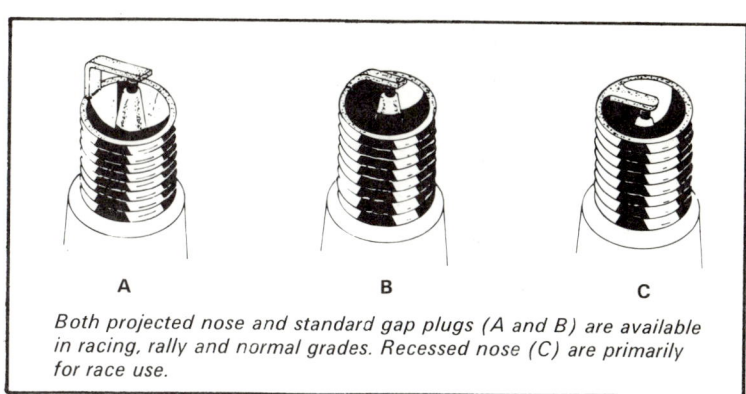

Both projected nose and standard gap plugs (A and B) are available in racing, rally and normal grades. Recessed nose (C) are primarily for race use.

known too many instances of these types causing an engine to misfire due to a faulty contact.

High tension (HT) leads

One of the most annoying faults on a standard engine is an incurable, irregular tick-over which on transverse engine set-ups often causes idler-gear knock.

Very often the problem lies in the fact that nearly all modern cars utilise what is known as TV and radio suppressed HT cable for plug and coil leads. The leads are composed of fibrous, high carbon content material. Unfortunately, if one measures the resistance to electrical conduction of such leads, then quite often one can find as much as 50 per cent variation between the leads on one engine causing considerable ignition variation between the cylinders. Fortunately the remedy is simple: replace all such leads with the old-fashioned but nonetheless more efficient continuous copper core cable. Variations on such leads are almost immeasurable excepting through the most sophisticated and delicate equipment.

Furthermore, the copper core is a better conductor anyway, thus giving a stronger spark at the plug, particularly on a full-race engine.

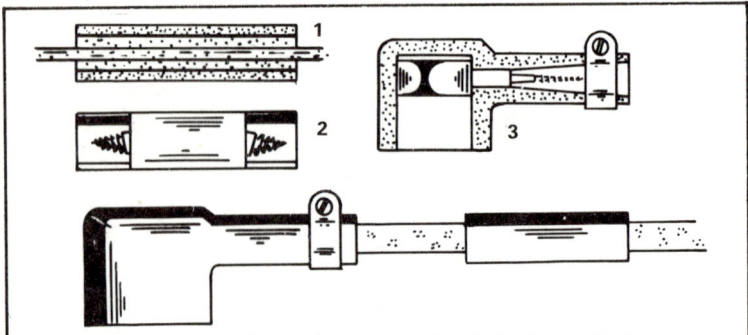

The copper wire strand runs through two insulating layers in the continuous core HT lead (1). Two screw points are located in either end of the suppressor (2) to screw into the copper core. This is again found in the rubber plug cap (3), linking the core to the plug connector.
A made up lead is shown below them, with cap and suppressor.

The only snag is that the law dictates that cars are fully suppressed and it thus behoves one to fit separate suppressors to each and every HT lead. I have never known these to give any trouble when in good condition but they should be replaced about every 18 months to two years.

Plug caps should be of the snug-fitting rubber variety, such as those coveted by the motor-cycle brigade and should be of the type that screw into the copper core of the leads. These caps are not only waterproof but being a tighter fit, they make better contacts with the plugs. They should be replaced every three years or so, since they do perish.

Generator
Needless to tell, rallying enthusiasts invariably fit alternators instead of dynamos, the former producing more current as demanded by the rally cars' battery of additional lights. However, not only are these very expensive but they do absorb up to 30 per cent more engine power. As anyone who has studied physics and the laws relating to

Mind the straw bales!

energy conversion will know, you can only take out of a system as much as you put into it, unless the system is to be allowed to run down (flat battery).

In this case the engine develops power in the form of mechanical energy which it derives from the chemical energy stored in petrol or whatever fuel it uses. Part of this energy goes to drive the wheels and part goes into the electrical system—ignore energy losses for the moment—to supply various electrical needs, including the lights. This latter energy is fed into the generator where it is converted to electrical energy. The greater the capacity of a generator for converting mechanical energy to electrical energy, the greater the amount of mechanical energy it will need to enable it to function. Hence the greater will be the amount of energy or power absorbed from the engine's output and less will be available to drive the road wheels.

Since each conversion stage involves some energy loss, usually as heat—dynamos get very hot and wires and lights all give off varying degrees of heat—one can begin to appreciate the magnitude of the problem. If one can reduce a generator's capacity or need for converting energy, then it figures that one can reduce the energy that it will absorb.

This is really only worthwhile on the most highly developed racing units and even then, gains are very small. However before rushing out to alter your generator, make sure, if you race that the regs allow such modifications. Mini-7 regs certainly don't. Even more extreme, one can, if regs allow, completely remove the generator, though I would only recommend this for short sprint/hillclimb type events. On longer events, reduction in battery output towards the end of the day may have adverse effects on ignition and more than nullify any small gains from dispensing with the generator.

With modern rev limits being much higher than was originally intended for the standard dynamo, there is a tendency for them to burn themselves out as a result of sustained high revs. Dynamo speed

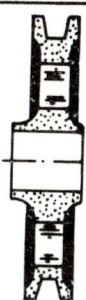

Holes of ¾" diameter can be drilled in the large dynamo pulley as shown (left) and for even less weight, metal can be removed from the shaded areas (right) to a depth of 20 to 25 thou and the component polished.

can be reduced using a larger dynamo pulley obtainable from BLMC Special Tuning (part no C-AEA 535) and it will necessitate the use of a longer fan belt (part no C-AEA 756).

These reduce the risk of dynamo burning and give a small bonus in engine power available to the wheels in that, through slowing the dynamo, they reduce its output and hence the amount of energy absorbed from the engine. This is only worth considering on a racing engine as the percentage increase is very small.

The pulleys, by the way, do have one fault. They are very, very heavy and can, with benefit to both overall vehicle weight and load on dynamo bearings, be lightened.

Distributor
One of the most confusing aspects of engine modification is the problem of distributors. They all look alike but are oh so different internally, having differing balance springs and advance plates giving differing degrees of inbuilt advance. So far as we are concerned we can regard the types as being split into four. Within these four types there are small differences but they are such as to make interchange-

ability of types a waste of time. As we are concerned only with interchangeability our four broad divisions hold good. The types are as follows:

Mini van and commercial vehicle types:
These are designed for use with the cheapest grades of petrol, so they have only limited inbuilt advance as a result. Consequently, provided one is prepared to use a higher grade of petrol, worthwhile improvements can be made simply by changing to the unit fitted to the equivalent cars. Often, superior fuel consumption offsets any increase in costs. Thus, for example, it pays to change a Mini van unit for an 850 Mini saloon type.

All other non-commercial vehicle types, excluding 997 Cooper and Cooper S:
This type with minor mods depending on the exact vehicle in question, covers the bulk of the A-series engines. Unless mods to the engine involve a change in camshaft there is little point in changing the distributor. If, however, a more vigorous camshaft is to be fitted then one of the following units should be used, assuming of course that the cam in question is of BLMC origin. If on the other hand a specialist regrind cam is to be used I suggest that you seek advice from the supplier. It may be that one of the following distributors will still be suitable.

997 Cooper and competition type No C-27H 7766:
Basically both of these have been designed for use with the 2A948 (now 88G229) cam which was fitted as standard to the 997cc Cooper. However the competition distributor is suitable for use with the 731 cam or even the early full-race 544 cam.

Cooper S type:
I can say without fear of contradiction that this unit represents the universally suitable distributor, provided that premium fuel is used. It is OK from the standard 850 to full race Cooper S and if ever I

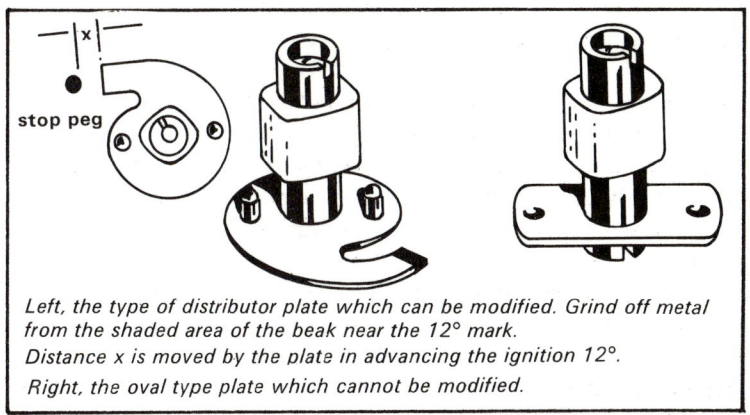

Left, the type of distributor plate which can be modified. Grind off metal from the shaded area of the beak near the 12° mark.
Distance x is moved by the plate in advancing the ignition 12°.
Right, the oval type plate which cannot be modified.

had to buy a new unit or modify an existing unit then this is the distributor I would use, unless I was unable to obtain suitable parts or already possessed a suitable alternative.

The distributors are all interchangeable without further mods to make them fit, though the Cooper S and competition units have no vacuum advance/retard mechanism. In fact this represents a worthwhile modification even on a standard engine. Disconnect the vacuum advance to prevent the possibility of over-advance and resultant bottom-end engine problems. Remember to bung-up the take-off hole in the manifold or carburetter choke tube. Another useful mod to any distributor is to fit a Cooper S contact-breaker kit. The points are a better material and have stronger springs which are useful if high rpm are envisaged.

I should point out that even on exchange, distributors are quite expensive and as an outright purchase they cost a bomb, the S unit being priced at £10 or so. However, do not despair, in nearly all cases there is, with a little effort, a cheap way out. I will deal with the exception later. It is possible, at a cost of only a few shillings, to modify even commercial vehicle distributors to full race Cooper S specification, or for that matter, any intermediate specification, assuming we regard the Cooper S as the ultimate.

First, dismantle your own unit and inspect the plate attached to the bottom of the distributor's cam rotor. This is the plate to which the springs and balance weights are attached. On the upper surface of this plate, by the beak, you will see stamped the maximum inbuilt mechanical advance of the distributor in degrees. Say that in your case this is 12 degrees.

If you partially dismantle the distributor you will find that after the top plate, to which the contact breakers, condenser, etc., are fixed, is removed, another beaked plate will be exposed, this plate being an integral part of the distributor cam. Measure very carefully the distance between the leading edge of the beak and the stop peg. Say this was Xmm. Divide this by the number of degrees stamped on the beak which gives an amount $X/12$mm. Since X is the distance moved by the beaked plate in advancing the ignition 12 degrees, then $X/12$ will be the amount of movement needed to advance the ignition by 1 degree. In standard form, once your distributor has advanced 12 degrees, the beak comes up against the stop peg thus preventing further mechanical advance.

If, however, you were to grind $X/12$mm off the end of the beak the amount of movement would increase by this amount and the advance would be increased from 12 to 13 degrees. Thus it now becomes a simple matter to grind the appropriate amount from the beak to increase the amount of inbuilt advance to whatever you require. Simply find out what figure in degrees is stamped on the ideal unit, subtract the figure stamped on your unit, multiply the difference by $X/12$ and you have the amount to be ground off the beak to achieve the extra advance needed.

The difference will rarely be more than 3 to 5 degrees. So if you ask for the degree advance of a distributor, remember that it's the degree of inbuilt mechanical advance that you want (usually about 10 to

John Rhodes in practice for the 1968 British Grand Prix meeting.

15 degrees) not the overall advance where a vacuum advance is fitted nor the degree advance required by the engine for which it is designed. On a racing engine this figure could be as high as 35 degrees.

Finish the job off by fitting the balance weight springs from the distributor, the specification of which you are aiming at. Disconnect and completely remove the vacuum advance mechanism, bellows and all having first joined the upper and lower halves of the top plate using solder and/or self-tapping screws. It is not strictly necessary to remove this mechanism but it makes things look much neater, particularly if the resultant hole in the side of the distributor is neatly and carefully filled with glassfibre or some other epoxy-resin filler.

When finished you have, for just about 10s, built yourself the equivalent of a Cooper S distributor or any other special type that you wanted. I said that this could be done with any of the distributors, van, saloon or what-have-you. This is true with one exception. One type of distributor which was fitted to the earliest Minis, 1959 and early 1960, had different internals to those just described and cannot be modified, it can only be replaced. It can be recognised by the absence of a beaked plate attached to the distributor cam. In this case the plate is much smaller and more oval. On the Mini, at least, its part number was 2A 995.

chapter eight

Tuning for economy

It seems to be a common misconception amongst the unenlightened that tuning an engine for speed will have an adverse effect on fuel consumption.

This is not necessarily true of a road conversion. In fact I would go so far as saying that if a normal road conversion causes an increase in fuel consumption, there is something wrong with the conversion. There are, of course, some exceptions to this rule; I understand that the 1800 BLMC B series engine is such an exception. Further I have always noticed an improvement in fuel consumption when modifying an engine from standard. A Mini Traveller I owned did 35mpg when driven really hard in standard form. Modified, my average consumption was 42mpg. Gentle driving did not show any improvement worth bothering about (2mpg).

There are other popular misconceptions. Probably the most common is the idea that the fitting of a larger but otherwise similar carburetter reduces economy, or that twin carbs give increased consumption. This again generally speaking is untrue, unless a carburetter change involves a change of type to a possibly uneconomical make (say change SU's for Amals). I changed my $1\frac{1}{4}$in SU for a $1\frac{1}{2}$in SU and found absolutely no difference other than an increase in performance (see end). This, of course, assumes that the carburetter is set-up properly, also that it is neither too large nor too small for the engine in question. When I first fitted the $1\frac{1}{2}$in carb I was down to 25mpg, but two days' feverish sorting (I had to get things right quickly or go broke!) gave me 42mpg (see end) with no reduction in performance. It is this question of fuel consumption which often distinguishes the really good tuning company from the 'cheap graunchers'. I am not knocking the inexpensive conversions or companies, just the 'graunchers'.

The point is this. When you tune an engine for extra performance, generally speaking you merely improve its efficiency. You increase the efficiency with which fuel mixture enters a cylinder, improve the efficiency with which it is burned (it burns more completely) and you improve the efficiency with which exhaust gases are removed.

This must, all things being equal, give better fuel economy. You may ask why a race-tuned engine is often very uneconomical. Without going into great detail, as outright economy is unimportant in racing and race-tuned engines are impractical on the road, this heavy fuel consumption is due to several factors but basically boils down to the fact that most racing engines use more fuel than they actually burn. A major factor in this is camshaft design, and the desire to have a cylinder filled only with fuel mixture at the point of ignition. Carburetter blow back on lift off etc, is a contributory factor.

Let us take another look at ordinary road tuning. How would I tune for economy and how would this compare with my tuning for speed? Well, firstly I will say this. There would be no basic differences. An efficiently gas-flowed cylinder head with raised compression is essential. Inlet and exhaust manifolds must be gas-flowed or replaced by free-flow 'one off' units. If my gas-flowing involved fitting larger valves, then if this necessitated fitting a larger carburetter, I would fit one. If, on the other hand, one can carry out worthwhile gas-flowing of the cylinder head without increasing valve size and/or carburetter size, do so, if economy is the sole aim.

The main differences, in fact the only differences, lie in the actual carburetter settings. Tuning for performance demands a slightly rich mixture setting. Tuning for economy demands a slightly weak mixture. Needless to say, all other operations to keep an engine working with the greatest efficiency, such as correct plugs, points, and tappet settings, reduction of engine friction etc, are common to both types of tuning. So there we have it. As far as I am concerned there are no differences, and I most certainly would not use a weak mixture to obtain slightly better economy; this tends to burn exhaust valves. One may improve economy by fitting a lower final drive ratio to raise overall gearing, but the reverse effect can be the case, if by doing so the vehicle becomes overgeared.

Tuning my Mini Traveller

I am often asked by various people what type of performance can be obtained by such and such conversion, and what degree of reliability, for simple everyday road use.

The ensuing section states briefly my personal experiences over 18 months, with my own everyday road transport, and should be typical of any converted road car. The vehicle in question is a February, 1966 Austin Mini-Traveller, with all-steel bodywork. In 18 months it covered some 40 thousands miles or so—I cannot be too certain as the mileometer was out of order for three months. I took delivery of the Traveller in completely standard form save Jaguar 3.8 Mk II headlamps with 70 watt main beams: a most worthwhile alteration.

The car was run in for four thousand miles on a standard head, then a spare head was fitted and work started modifying the 'new' 850 head. It was fully treated in accordance with my earlier Mini tuning book, 100 thou were taken off the face and Morris 1100 (not the larger MG 1100) valves were used on the inlets. Standard exhaust valves were retained. Single extra strong valve springs were used though I did not fit any steel shims under the collets. The combustion chambers were carefully balanced, the manifolds gas-flowed and matched to the head. Standard valve guides, re-shaped, were retained.

107

Initially, with economy of expenditure, in mind, I stuck to the
original HS2 1¼in SU carb.
On assembly I found that although performance was much better
it still seemed strangled. The 0-60 time was approximately 19 seconds.
I set to work on the carb, fitting quicklift dash-pot assembly and
E3 needle. This reduced the 0-60mph time to 17 seconds, still not
good enough, even though a Traveller is 1cwt heavier than the
ordinary saloon. I fitted a No 6 needle. Acceleration time dropped to
15 seconds for 0-60 and the top speed was approximately 85-90mph
(fifth wheel reading on a short straight of 88mph) but fuel economy
was disastrous, 26mpg! Bad enough to break the bank. Far too rich,
it even cut out at the top end.

Well, I tried radiusing the piston and belling out the carburetter
mouth, but all to no avail. Finally I admitted defeat and fitted a
secondhand 1½in H4 SU on a suitably modified inlet manifold.
A quick-lift dashpot assembly was combined with an RLB needle
and it was absolutely fabulous. With only head and carb modifications,
I could see off any ordinary Cooper and quite a few Cooper S types,
including my mate's standard 1275 Hydrolastic S. We obtained 0-60
in 12 seconds (we have often been accused of over-optimism on this
figure) which we checked and double checked on several watches and
top speed on the flat was in excess of 90mph; in fact under ideal
conditions an indicated 105mph could be obtained! The governing
factor was valve-bounce in top! Petrol economy was a very acceptable
42mpg. Oh yes, two other mods. Michelin X tyres and lightened
rockers and push-rods. All brakes and suspension were standard.
Although the former were more than adequate I am afraid that the
same cannot be said for the suspension. It was horrible! Talk about
rock and roll. The net result was three 'done' drive shafts within
three months. I was accused of brutal driving by my garage (partially
true) to which my reply was as always a few oaths directed at the
shock absorbers as fitted to Minis.
A pair of Konis was fitted to the front only, set on soft settings.
Not only did these transform the former somewhat 'ship-at-sea'

type of ride, but it also cured the drive-shaft unreliability, no doubt by preventing excessive front suspension travel. These were fitted at 11,000 miles; on selling the car at 40-odd thousand miles no further drive shafts were needed. With the number of warranty claims made on drive shafts, particularly the rubber-cross UJ couplings, I wonder BLMC don't get wise and at least use a decent specification shock absorber on the front.

I don't blame the manufacturers of the shockers as they are undoubtedly tied to a price and given specification. Perhaps Granny does not get much trouble, but many others do. The brakes were re-lined on the front at 25 thousand miles with Ferodo AM4 and these were only half worn when the car was sold. Oh yes one small mod. I forgot. At the same time that I fitted the shockers I fitted an SPQR Mini engine stabiliser kit, consisting of small cones and rods under the gearbox. Not only did these stabilise the engine, but road-holding was further improved by cutting out engine/transmission rock which transfers to the front wheels.

Other than the usual Mini mods of an SPQR throttle cable and steering column adjuster, and straight-through Sprite Mk I silencer, no other deviations from standard were made.

Really and truly this was a fabulous little bomb. The performance was quite surprising, to judge by people's faces when they were passed, being superior to a Cooper and at least equal to Cortina GT. Fuel economy was great and after 40-odd thousand, 400 miles still only needed one pint of Duckhams Q20/50 oil. This latter is in my estimation quite the finest oil I have ever used in a road car, and although I am always willing to be convinced otherwise, I would use none other. It retained its pressure under extreme conditions (65-70psi) and did not sludge.

In the last three months of my ownership of the Traveller I used it for towing my Mini 7 Racer and this it did very well in the dry, being capable of 70 plus mph and 35mpg. Some 3000 miles were covered under these conditions. However there were two big snags when towing. You just could not stop under any circumstances, and anything, even a straight line, was lethal if it rained. After spinning the whole lot twice at 15mph. I decided to trade the old faithful in for a new 1500 Cortina Super—more about this in the future.

The only expenses other than petrol and oil and the aforementioned drive shaft, etc had been five new Michelin X's—two more needed at the time of sale—new wheel cylinder rubbers, one new exhaust system, two fan belts, eight sets of Champion N9Y plugs, three sets of points and a new clutch thrust bearing which I never did fit, it had been on the way for several months and finally went kaput the day I parted with the car. All in all one of the most reliable and faithful cars I have ever had or am likely to have.

chapter nine

Modifications for alcohol fuel

Although the contents of this chapter refer to a 1293cc Austin
Cooper S, the principles involved and problems encountered are the
same as for almost any other standard production motor car.
While there may be some of you who do not agree with our approach
to the problem, I can only say that the exercise was successful and
our objectives were achieved as will be seen at the end of this chapter.
The first thing to realise is that when using alcohol fuel the amount
by volume that needs to be burned to obtain a given pressure on
the top of the piston is greatly increased when compared with petrol.
Alcohol has an air to fuel mixture ratio of 8:1, whereas this ratio
is 14:1 with petrol. Secondly, the heat released by alcohol when
ignited is far less than with petrol, with causes a certain amount of
difficulty in obtaining an efficient working temperature. Thirdly,
one can use a much higher compression ratio to good advantage,
the limits being set by one's engine strength rather than the fuel's
anti-knock properties.

Finally, vegetable castor based oils are extremely soluble in alcohol
and, if used, would immediately be washed away from the cylinder
walls, causing undue bore wear and piston ring breakage or seizure.

Our approach to the problem
The first thing was to decide on what fuel to use and we finally
decided on 90% alcohol, 5% acetone and 5% petrol; then we had to
find a supplier. After telephoning various fuel companies with little

success, we were finally put in touch with a supplier in Croydon and a stock of fuel was obtained.

Turning next to the lubricant. We were rather fortunate in being able to obtain a supply of Esso Synthetic Grand Prix oil, exactly as used by Jim Clark that year in his World Championship-winning Lotus-Climax. As we had been running on Castrol R, we were forced to completely strip the engine and gearbox, wash it thoroughly in meths and surgical spirit and then re-assemble it using the Esso oil as our lubricant. This was essential to prevent any remaining Castrol R film reacting with the new synthetic oil, and causing blocked oilways, with obvious dire consequences.

We decided that a 14:1 compression was about as high as the engine would stand without breaking pistons and continually running big-ends bearings. A new cylinder head was obtained and gas-flowed exactly as before, but attention was paid to obtaining a very high polish as this prevents the head from absorbing too much heat for as already stated, alcohol fuel poses certain problems when it comes to obtaining a suitably high working temperature. To this end, we also removed the fan and blanked off the front grill, leaving only sufficient air intake necessary to feed the carburetters, the net result being a working temperature such as was obtained on petrol.

This took care of the simplest and most easily carried out modifications, the camshaft, pistons, manifolds and carburetter types remaining exactly as before. However, the major and most time consuming operation still remained; obtaining the right mixture and fuel flow! The mixture ratio of 8:1 means that when using alcohol, one must considerably increase one's rate of fuel flow by as much as 150 to 200% if starvation and/or a weak mixture are to be avoided, though conversely, if the mixture is made too rich then oil dilution and bore scouring becomes a problem.

Several days were spent over pints of beer and cups of coffee deciding how best to approach the problem, without too much expense, or too much completely new development work, using different carburetter

112

types (we were using twin H4 1½in SU's) such as Webers or Amals. Finally we came to the following decision. Our fuel system should be capable of pumping 30 gallons of fuel, into the inlet manifold, per hour, if the needles were removed. Thus we would obtain the required percentage fuel flow increase, but rely solely on the carburetter needles to give the correct metering under working conditions. We decided to start at the fuel tank and work right through the system step by step, finally dealing with the carburetters and needles.

The tank capacity was first considered. Fuel consumption on petrol had been about 11mpg giving a range of 50-55 miles on our existing carrying capacity. Consumption was calculated as being about 3mpg on alcohol (we were miles out in this as will be seen later) giving a range of 15 miles. Since our objective was the Brighton Speed Trial run over a standing start kilometre we felt that our existing tank capacity was sufficient and this proved to be correct.

Due to more fuel being taken out of the tank per hour, more air obviously had to enter to take its place, or a vacuum would develop preventing further fuel flow. Thus the filler cap had a hole ⅛in diameter bored in it to allow for any extra air intake.

The existing competition Cooper fuel pump, manufactured by SU, was tested and found to deliver approx 12 gallons of fuel per hour to the carburetters. This was insufficient as we wanted 30 gallons per hour, but at the same time did not wish to increase our supply pressure to a point where the float chamber needle valve was unable to make a perfect seal, which would have caused carburetter flooding.

Confronted with this problem, we approached SU's who supplied a double pump which met our requirements and delivered approximately 30 gallons per hour. We then decided that although the pump and tank capacity were sufficient, the internal diameter of the standard fuel line would be such as to restrict the fuel flow rate beneath the pump's capabilities. This problem was overcome by replacing the standard fuel pipes with $\frac{7}{16}$ bore, alcohol resistant plastic piping. This was re-routed along the car so that it ran inside and not underneath the body shell. Tests were made and we found that the flow rate was approximately 30 gallons per hour with little pressure increase. Thus all that remained was the job of modifying the carbs to give the right mixture and herein lay the difficulties.

Further calculations were made and we decided to increase the jet size from .090 to .125, and further findings showed that as far as we were concerned the easiest method was to bore the existing .090 jets to this size. Great care and extremely accurate work was obviously the order of the day. Fortunately no snags were encountered and the operation was completed without hitch. Of course it was no use increasing the jet size without also increasing the delivery capacity of the needle valve and jet. T4 needle valves were suitably modified and the bore sizes increased by 100%

Further tests were carried out and the flow rate was found to be 27 gallons per hour through the main carburetter jet, in the absence of a needle. The float chamber needle valves did not protest and flooding did not occur.

So far so good, but now we had to meter the fuel which we were quite effectively pushing into the carburetters. nb—Remember the way that the needle meters the fuel flow. The needle is tapered, the thickest ending being .001 smaller diameter than the jet, eg .124 in a .125 jet. Thus the thicker the part of the needle within the jet, the less is the amount of fuel allowed to flow through the jet due to the obstruction caused by the needle. Conversely the thinner the needle within the jet the more is the fuel allowed to flow. Thus thick needles give weak mixtures, thin needles give rich mixture.

It follows, therefore, that less fuel is needed with small throttle openings and hence the thickest part of the needle will be within the

114

jet, whilst at maximum rpm most fuel is needed and hence the thinnest part of the needle should be within the jet.

It can be seen, therefore, that if a .090 needle, .089 at its thickest point, is fitted to a .125 jet it has the effect, at say 1000rpm, of allowing enough fuel for say 4000rpm into the cylinders. This, of course, is flooding the engine. In fact a 125 needle has a diameter of approximately .089 a third the way down its taper.

Firstly we had to find out what was the total lift of the needle, within the jet, under full throttle. Rightly or wrongly we worked on the basis of measuring the amount of lift that the dimensions of the carburetter would allow. This was taken as representing the lift at full throttle. Also we needed to know how much of the needle projected into the jet when the engine was not running, and also when on full throttle. This was done quite simply using a depth gauge inserted into the piston damper, giving us the amount of lift,

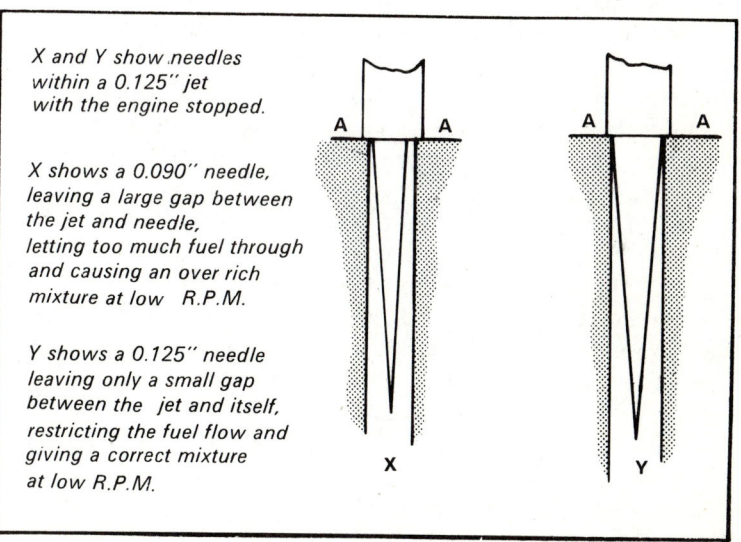

X and Y show needles within a 0.125″ jet with the engine stopped.

X shows a 0.090″ needle, leaving a large gap between the jet and needle, letting too much fuel through and causing an over rich mixture at low R.P.M.

Y shows a 0.125″ needle leaving only a small gap between the jet and itself, restricting the fuel flow and giving a correct mixture at low R.P.M.

then measuring the length of needle projecting from the piston and the depth of the jet below the piston bridge. This gave us the amount of needle in the jet at full lift and with the engine switched off. From this we were able to see over which range of the needle taper the carburetter operated. All this was carried out using the jet needle set-up used for petrol. On these findings we based our next calculations and modifications, since we assumed that the basic engine characteristics would not be altered, rather they would be accentuated and moved up or down the rpm scale.

Since the new set-up involved the use of a .125 jet we were set a serious problem. Our existing needles, though possibly OK when the throttle was fully open and only the extreme tips projected into the jet. were designed for use with a .090 jet. Thus at very low rpm the mixture would have been incredibly rich, and the effects would have been most unpleasant.

The obvious solution appeared to be to use needles suited to .125 jets. Unfortunately our set-up involved the use of bored out .090 jets (.125 jets are for use in 2in SU's) as we were using 1½in SU's. These .090 jets and needles are much shorter than the normal ·125 jet which therefore naturally enough, utilises longer needles. We also decided that it was impossible for us to make our own special needles.

After much thought another idea was born. We would start off with standard .125 needles, reduce the length to that of our existing .090 BG needles by cutting off the tip at the thin end. This was duly done, and the services of a friendly watch-maker were sought. He kindly agreed to turn the shortened needles down to our own profile, which we had already calculated. In actual fact 3 sets of needles were made up this way, one set being weaker, one set richer and one set about right, according to our calculations. Thus enabling us to alter the mixture as we saw fit.
While these needles were being made up we decided to try one other method of making up a suitable needle. Standard .125 needles were

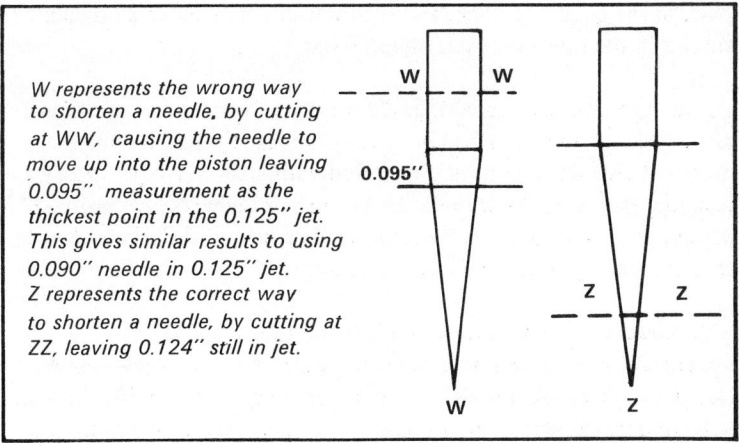

W represents the wrong way to shorten a needle, by cutting at WW, causing the needle to move up into the piston leaving 0.095" measurement as the thickest point in the 0.125" jet. This gives similar results to using 0.090" needle in 0.125" jet. Z represents the correct way to shorten a needle, by cutting at ZZ, leaving 0.124" still in jet.

shortened by cutting them off at the thick end, thus effectively moving them up in the jet. This, of course, is what happens when the throttle is opened, causing more petrol to enter the choke tube. At low rpm, however, this over richens the mixture. The results were disastrous, fouled plugs and very much diluted oil, so much diluted that it was easily ignited by a match. The most unfortunate effect, however, was that the alcohol 'in' the oil attacked the crankshaft oil seals, these broke up and let the oil into the clutch!

In fairness to ourselves, however, I should point out that this was not entirely unexpected and it was rather in the nature of a 'suck it and see' experiment.

After a few days, during which time we stripped and rebuilt the engine, fitted new oil seals and clutch etc, the new specially made needles arrived, and were hastily fitted.

The engine was warmed up and away we went. One short run was enough to convince us that we had the answer for short sprints. Both torque and bottom end power were very much increased. However, back to the garage we went and concentrated now on small detailed tuning in the quest for even more power.

At this time we discovered that due to the evaporation properties of alcohol the carburetters had been icing up, as also had the inlet manifold. We decided to fit a manifold with a shorter pipe, thus bringing the carbs closer to the engine and its higher temperature. This prevented further icing, but necessitated removing the piston dampers to correct the resultant mixture variation.

Still, however, the mixture was a little on the rich side, but fortunately right throughout the rpm range. Thus we knew that we had the right needle profile and since the mixture was only a little too rich we decided next to fit a reverse cone megaphone, assuming that this would have the effect of weakening the mixture, as well as giving more power. This proved to be correct on all counts and, 'Oh boy! did that vehicle fly!' 8,200rpm being easily obtainable in all gear s *including* top, still using the original final drive ratio. Top speed was not unduly increased, but acceleration was improved almost beyond belief, 125mph being achieved in the twinkling of an eye. Fuel consumption? Well, to put it mildly, very high $\frac{2}{3}$ of a mile per gallon!

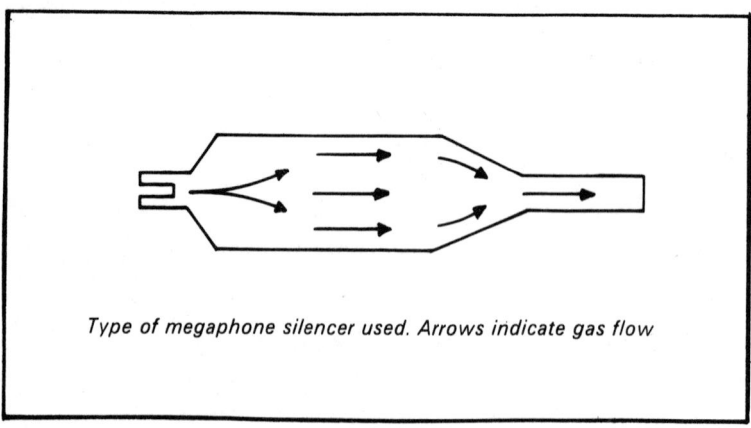

Type of megaphone silencer used. Arrows indicate gas flow

The Anglia waits for the Minis to spin and leave a gap.

One further problem arose; fumes. The fumes were intolerable, making one's eyes, and nose stream, giving a sore throat and making one cough. This was quickly and easily put right, however, using adhesive linen cloth as a sealer. How did we shape at the Brighton Speed Trials? Well, firstly, we were put in with all the sports/GT cars, which class included Lotuses, Porsches, etc. This because the organisers rightly claimed that we were highly modified.

Unfortunately we could have done better. The driver thought the first timed run was a practice run and switched the engine off, to coast the last 30yds over the finish! It rained for the second timed run! However, still the results were impressive. We were second fastest in class, being only .3 secs slower than the winning Lotus and being second fastest saloon present. A Cobra-engined Cortina (4.7 litres as opposed to 1.3) was just 2 hundredths of a second faster, the car in question belonging to one Doc Merfield.

Our time for the standing kilo was 26.48secs and our terminal velocity was approximately 115mph. This was well inside the existing class record and 7th fastest time of the day.

chapter ten

Non-Cooper S
1275cc units

This unit exists in two basic layout forms. The normal inline type as
fitted to Mk 4 Sprites and Mk 3 MG Midgets and the transverse
version as fitted to 1300 versions of the earlier BLMC 1100s.

On Sprites and Midgets twin $1\frac{1}{4}$in SUs are fitted but the cylinder
head only has nine studs. The 1300 GT saloon announced at the
September 1969 Motor Show has similar carburetters but in addition
has eleven cylinder head studs. The ordinary non-GT versions and the
1275 long-nosed Mini Clubman both utilise single carburetter units,
otherwise all these engines can be regarded as basically similar. Due
to the similarity to the Cooper S 1275 unit there exists a considerable
amount of confusion amongst enthusiasts as to just what are the exact
differences and advantages or otherwise of the two engine types
(Cooper S or non-Cooper S units). The following endeavours to
throw some light on the matter.

Naturally enough when this engine was first announced in its inline
form for the Sprite and Midget most people thought that this was
just a slightly modified S unit modded to suit the different layout.
However the trained eye soon spotted that there were only nine
cylinder head studs as compared to the eleven on the Cooper S. This,
however, is not all, variations from S specification are more than just
superficial. Though port and chamber shape of the heads are similar
valve sizes are not. Both inlet and exhaust are smaller and although
the exhaust valves are of racing quality being made in Nimonic

material, the inlet valves are not of such a good quality as found on the S head. As on the latter head, there is no oilway running across the head face about 120 thou. below it as on other A-series heads. Breaking through into the oilway when raising the compression by machining the head is a problem that has once again been eliminated.

Needless to say since the first announcement much more has been learned, and the Mk 4 head, as this latest head has become known, has proven itself a valuable source of extra power on all racing Minis whether they be 850s or full-blown 1293cc Cooper Ss. The thing is that provided the exhaust valve size remains the same it is possible to fit even larger inlet valves than fitted to the Cooper S, it being impossible to increase sizes of inlet or exhaust on the Cooper S head.

Thus on Cooper S derived racers we find exhaust valves remaining the same as on the Sprite but inlet valves up to $\frac{1}{16}$in larger than those fitted to the Cooper S. On 850 and 998 Mini racers the smaller exhaust valve of the Mk 4 head is a distinct advantage (compared to the larger S exhaust valve) whilst the inlet valves can be increased to the size fitted as standard to the Cooper S. This is larger than can be fitted to any of the more ordinary heads such as the 12G295 casting which is fitted to MG 1100s and 998 Coopers. The Cooper S head is far from ideal for the smaller units (excluding 970 S) since exhaust valves are too large. On racing 1293 S units the smaller exhaust has proven quite adequate though not beneficial other than that it allows the use of much larger inlets. For these reasons the Mk 4 head had emerged as the superior head for all racing applications.

It is of course necessary to drill the nine studded versions to convert to eleven stud fitting as used on the Cooper S and the 1300 GT, the head on the latter, and for that matter all non-S 1300 units, being as described—all can be considered as Mk 4 heads. One problem arose when the Mk 4 head was first played with. It was discovered that it was somewhat thicker than any other head measured from the valve seat to the bottom of the valve spring locating groove. In the early days this caused all sorts of problems and some people had special longer inlet valves made to prevent spring crush occuring àt full valve lift, the exhaust valves were alright since they remained the standard Mk 4 type and were designed for the head in the first place. Others, like myself, who did not want larger than S inlets, chose to do away with all of the Mk 4 valves and fit Cooper S exhaust turned down to a suitable size and standard Cooper S inlet valves. To prevent spring crush we fitted the shorter but equally 'strong' Terry extra strong Cortina 1500 GT valve springs. Now, however, special valves are available through BLMC Special Tuning at a reasonable price. These valves are suited to the Mk 4 head and come either in standard Mk 4 or standard Cooper S sizes. Unfortunately, though this makes valves available to 850 and 998 Mini tuners at reasonable prices it does not help the chap who wants to fit really large oversize valves and he still has to buy expensive special valves from the specialist tuning companies, unless he can use something like Ford racing valves as blanks and turn them to size.

The Mk 4 head is only fitted with the old iron guides and it is necessary to replace these with hidural bronze types such as are now only available through specialist companies such as Janspeed or Downton Engineering.

Since the Mk 4 head is designed for an engine with the same bore centres as found on the Cooper S and since these differ from the more mundane A-series types (850, 948, 1100, 998, etc) this head does not represent other than an outright racing tweak. To overcome the reduction in squish area resulting from fitting the later head to the 850 type units and its resultant bad effects on combustion chamber turbulence and combustion characteristics, it is necessary to use very high compression ratios in the region of 13 to 1, and this is not in my opinion acceptable other than on a racing unit. These problems do not, of course, arise on the S unit or the units for which the head was designed in the first place.

Looking at the block one finds that the 1300 unit does not have any inspection plates over the camshaft-follower chest, the block being solid in this region. This is supposedly to increase block rigidity, though I am inclined to think that more likely it is just somewhat cheaper to make it that way. If rigidity is increased it is a hindrance to reliability (discussed in a little while). Because there are no removable plates it is damned difficult to overhaul or work on this part of the valve gear.

Real increases in rigidity are to be found around the main bearing housings and the flange which bolts up to the gearbox casing. The housings are much stronger and stiffer and the flange is about $2\frac{1}{2}$ times thicker than on the S and some other A-series units are now appearing with the thicker flange only. This thickening of the flange was an effort to prevent movement of the block relative to the gearbox causing the sump gasket to leak and oil to spill out on to the ground. Unfortunately the engine now seems to have become so rigid that instead of flexing as the 1275 S block does if fully race tuned to 1300cc, the block not being able to flex or give a little just splits. I have seen several blocks of this later type with great cracks formed around the dipstick hole. One authority maintains that this is purely because people overlighten their flywheels and thereby lose its torsional vibration damping effect. Whilst I do agree that for this reason it is possible to overlighten flywheels it is strange that the same degree of lightening causes no problems on the S blocks. I have a friend who uses a full-race unit from this particular tuner and his flywheels are in fact lighter than any A-series unit I have ever seen

Connecting rods are rather similar but much heavier than on the S unit having large counter-balance weights on the big-end caps in an effort to overcome the roughness with which the 1300 unit seems to be endowed unless race prepared.

Crankshafts are Tuftrided rather than nitrided as on the S, though I understand that in the interests of economy many of the latest S crankshafts are also Tuftrided, and in this case both S and 1300 cranks would be identical (bearing in mind of course variations for layout on the inline units). Due to lower compression ratios 1300 pistons have much larger dishes in them than are found on the Cooper S units. As covered in an earlier chapter oil-pump-drive and camshaft tails show a variation but I have already fully covered this in that chapter.

The foregoing differences excepted, one can consider the 1300 and Cooper S units as pretty well identical. Do remember that if you are tempted to fit a 1275 Cooper S block to a Sprite or Midget that it will be necessary to make a complete new rear main bearing cap as those on the inline engine differ in shape to those on the transverse engine.

MINI SPORT for the LARGEST STOCK of
NEW and USED MINI SPARES and
ACCESSORIES in the NORTH.
(with a money back guarantee)

OIL COOLERS:

13 row oil coolers complete with pipes and block connections: to suit Mini, Cooper, 'S' etc, OUR PRICE only **£9 0s 0d**

MAMBA LIGHT ALLOY ROAD WHEELS:

Available in 5" and 6"×10". Natural Aluminium and Deluxe finishes.
Natural finish (both sizes) OUR PRICE only **£27 0s 0d** set of 4 (or **£6 15s 0d** each) (Usual retail £7 9s 6d each)
Deluxe finish (both sizes) OUR PRICE only **£31 0s 0d** set of 4 (or **£7 15s 0d** each) (Usual retail £8 15s 6d each)
NB—Deluxe finish also available in a wide range of colours including: Yellow, Red, Blue, Green OUR PRICE **£7 15s 0d** each.

PERSPEX WINDOWS

Available in clear or tinted, to suit Mini etc.
OUR PRICE only **£13 10s 0d** complete set (less front screen)
(Tinted Blue/Black **£3 0s 0d** extra per set).
Also available separately—prices on application

STEEL ROAD WHEELS:

4½×10 Mini, Cooper, 'S'	only **59/6d** each
4½×10 All Minis—	
Reversed Rim	only **62/6d** each
5×12 to suit Mini etc	only **59/6d** each

MODIFIED CYLINDER HEADS:

850 Mini Stage II cylinder head. Compression raised to between 9.5 and 9.75:1. Polished ports and combustion chambers and stronger valve springs. Ex. price **£13.10.0d.**
850 Mini Stage III cylinder head. Complete as Stage II but with new larger inlet valves. Ex. price **£16.10.0d.**

CAMSHAFTS:

Reprofiled, hardened and parkerised. Available as follows:
BMC 649, 731, 544 **£8.10.0d.** ex. **£9.10.0d.** outright.

PIPER CAMS:

As we are agents for Piper Cams we always have in stock their range of camshafts. Please enquire for further details.

½ ENGINES . . . BRAND NEW:

The following ½ engines etc are covered with a 12 month or 12,000 miles warranty:

998 Cooper ½ engine	**£42 0s 0d**
1275 'S' ½ engine (genuine)	**£92 0s 0d**
970 'S' ½ engine	**£85 0s 0d**
MG 1275 ½ engine	**£69 0s 0d**
1275 Spridget ½ engine	**£69 0s 0d**
1275 'S' Block only	**£51 0s 0d**
970 'S' Block only	**£51 0s 0d**

NEGATIVE CAMBER KIT:

2° front 1° rear OUR PRICE only **£5 10s 0d** complete set
with adjustable rear OUR PRICE only **£7 10s 0d** complete set

ENGINES & GEARBOXES, etc.

We always have in stock a wide range of good quality used engines, gearboxes, blocks, cranks, cams, cylinder heads, etc, etc for Mini, Cooper and 'S'. For fully comprehensive details and up-to-date prices just send self addressed envelope for our 'SPARES LIST'.

MISCELLANEOUS PARTS:

Roll-Over Bar: only **£8 10s 0d**
Mini Sump Guard: only **£4 5s 0d**
Rally Seat Cover: only **£3 19s 0d**
Rally Jackets—all sizes only **£5 10s 0d**
Racing Overalls—all sizes only **£6 19s 0d**
'S' High Capacity Water Pump: **£3 0s 0d**
'S' Distributor: **£7 0s 0d**
'S' Extractor/Exhaust manifold: **£12 15s 0d**
Duplex timing gears and chain: **£5 10s 0d**
Steel Centre Main strap with longer HT bolts: **£3 0s 0d**
1300 Rocker Shafts complete: New **£6 0s 0d**
Right Hand Fuel Tank with fitting kit: **£14 0s 0d**

Ref 'A'
PLEASE NOTE: Items listed here represent only a small portion of our entire stock.
For full details of our vast and ever increasing range of spares and accessories send now for our present day 'SPARES LISTS'—free on request.
REMEMBER all items are covered with our money back guarantee.

GLASS FIBRE PANELS:

These panels are of top quality and yet you will notice, are some of the cheapest available.
Mini FRONT END (in BMC colours): **£13 10s 0d**
Mini FRONT END (in Primer): **£12 10s 0d**
Mini CLUBMAN FRONT END (in BMC colours): **£16 0s 0d**
Mini CLUBMAN FRONT END (in Primer): **£15 0s 0d**
Mini WHEEL SPATS (BMC colours and pattern):
standard size **£4 10s 0d**
large size **£5 10s 0d**
Mini BOOT LIDS (BMC colours): **59/–d**
Mini BONNETS (BMC colours) **£4 15s 0d**
Mini DOORS (Primer) with inner panel and window from **£8 0s 0d** each or **£15 0s 0d** pair
Mini REAR VALANCES (Primer)
to suit van **45/–d** each
to suit car **38/–d** each
NB—All prices shown are correct at the time of going to press. Carriage charges extra on all items.

MINI SPORT   (Dept MMT)
7 & 15 Church Street,
PADIHAM, Lancashire
Tel. 73285   (STD 0282)

STOP!
LIKE A CHAMPION
with *MINIFIN*

REC. RETAIL PRICE
£8·19·6
PER PAIR

* Increase braking efficiency
* Cast in Iron liner
* Fits all Standard BLM's with 7" drums

CROSSFLO ROCKER COVER

In Polished Aluminium
Increased cooling
Can be fitted or easily adapted to fit all standard BLM's 'A' series engines

Rec. Retail Price £4

For Further Details of Stockists and Trade Enquiries Write to THE SOLE MANUFACTURER :

J.V. MURCOTT & Sons Ltd.,
Grosvenor Road, Aston, Birmingham 6.
telephone : 021- 327 2671 / 2 / 3

TUNING EQUIPMENT FOR YOUR MINI

CYLINDER HEADS
stage 1, 11, 111

WEBER CARB KITS
28/36/DCD
40 DCOE   45 DCOE

STROMBERG CARB KITS
single . . . . .125 CD
150 CD
175 CD
twin . . . . 125 CD
150 CD

S.U. CARB KITS
single   1½
  ,,   1¾
twin   1¼
  ,,   1½

EXHAUST MANIFOLDS
four different types

CAMSHAFTS for
road, rally, race

ALLOY ROCKER BOXES

STEERING WHEELS
wood, leather

OFFICIAL WEBER DISTRIBUTORS
CONVERSION CENTRE LTD
45 TULSE HILL, LONDON SW2
TEL: 01-674 5014
Send 2/6d for our latest catalogue

Don't take chances!
INSIST ON
PECO BIG BORE silencers

Single Pipe✳
from **77/-**
Twin Pipe ✳
from **114/6**

As Britain's leading manufacturer of Performance Exhaust Systems, we are flattered that attempts are made to copy the outward appearance of our patented Big Bore Silencer.
BUT REMEMBER ONLY A GENUINE PECO BIG BORE SILENCER CAN OFFER :—
● UP TO 10% improvement in performance* as a result of the unique internal design (U.K. Pat. No. 1111473/4)
*e.g. Motor Road Test on MGB 1967.
● Longer service life due to copper plating of main internal parts and use of high grade materials.
● PECO Noise Test and Guarantee Certificate supplied only with genuine PECO Silencers.

Available from
HALFORDS
all good Garages and Accessory Shops. If in difficulty write to

Austin/Morris Mini and Mini Cooper Saloons.

Look for the **PECO** *trademark on each Silencer.*
✳ Correct at time of publication.

PECO SILENCERS LTD. BIRKENHEAD

Birkenhead Office:
051- 647 6041 (10 lines)
London Depot: 01- 947 0993

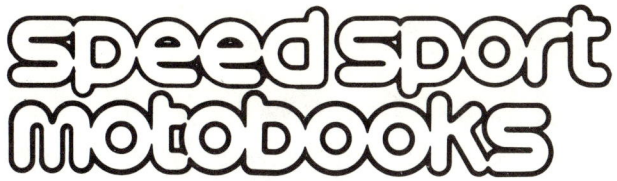

THE
PERFORMANCE
PAPERBACKS

are now published by

INTERAUTO

INTERAUTO BOOK COMPANY LTD

The following pages present to you some of the current SpeedSport and Interauto books for the motoring enthusiast, the automobile technician and the motorist.

If you have enjoyed this book you will find the two titles below of interest – from the Speedsport Motobook Range

THE THEORY & PRACTICE OF
HIGH SPEED DRIVING
Walter Honegger

English version by Charles Meisl

Hard-back bound. Profusely illustrated £1.50

A new book that demonstrates practically how the technique of your driving can be improved. All drivers can use this book to obtain higher average speeds and greater safety. Devoid of long winded theoretical debates, this is a handbook from which the racing enthusiast or road driver will derive immediate benefit.

The author is chief instructor and co-organiser of the Swiss International Racing Drivers Course.

THE THEORY & PRACTICE OF
CYLINDER HEAD
MODIFICATION
David Vizard

Hard-back bound. Profusely illustrated £2.50 (Paperback £2.00)

The definitive work for the professional but easily read by the novice.

Its contents include :
Workshop equipment, Valves, Guides and Seats, Matching of Ports to Manifolds. CR Calculation. Chamber Reshaping/Balancing methods. Port Design. Combustion Chamber Design. Optimum Valves, springs and plugs — plus specific details on most popular engines and actual size chamber modification templates.

Motorsport

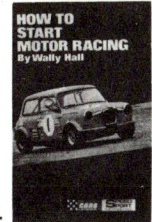

HOW TO START MOTOR RACING. Wally Hall. 011.9. £1.00

The author has had considerable club racing success and has passed on most of the vast experience he has gained. Ideal for anyone at all interested in beginning.

HOW TO START RALLYING. Colin Malkin. 024.0. £1.00

This famous rally driver takes the reader through all the mystiques of rally preparation. Car selection, suitability and setting up. Bodywork, lights, driving and navigation are some of the subjects dealt with. Colin co-drove the winning London to Sydney Marathon car.

HOW TO START AUTOCROSS AND RALLYCROSS. Peter Noad. 033.X. 80p

Like the rest of the 'How to Start' series but for the increasingly popular sport of autocross/rallycross. Like the other authors Peter Noad is an experienced and successful campaigner.

TOUCH WOOD. Duncan Hamilton. 042.9. £1.50

Paperback   25 b/w illustrations

We feel that we have found a great book in **TOUCH WOOD** and have re-issued it as the first title in our reprint series of motor racing classics. Duncan Hamilton was typical of the enthusiastic amateur who went racing for the sheer hell of it. He drove many makes of car to their limit, mostly Jaguars, utterly indifferent to his own safety and surviving many spectacular accidents. He won Le Mans at over 100 mph suffering from a monumental hangover, crashed an aeroplane, was torpedoed twice and helped to put England back on the motor racing map. His autobiography is a marvellous colourful story and has been out of print for a long time.

HOW TO START PRODUCTION MOTOR CYCLE RACING   Ray Knight   030.5   £1.00
Ray Knight

Ray Knight is a journalist with 10 years racing, a TT win and lap record to his credit. He passes all his experience to the enthusiast. 'A good guide to success.'

THE *BARRY LEE* BOOK OF HOT ROD RACING   062.3   £1.00

Barry Lee revolutionised Hot Rod Racing in 1970 and in 1971 became British Champion, as well as making successful forays to Denmark and South Africa. In his book Barry Lee shows how he built his Escort, what it's like in a hot rod race, where and when hot rod racing takes place - in fact he writes about everything that an intending competitor, a hot rod fan or spectator will want to know.

Marque tuning guides

TUNING THE MINI.   Clive Trickey   001.0. *£ 1.00*
The Mini Tuners' Bible; universally recognised as the most authoritative book on the subject. The most popular marque tuning book to be published.

MORE MINI TUNING.   Clive Trickey   000.3. *£ 1.00*
New! The second edition of the companion volume to **'Tuning the Mini'.** Updated with much more information on valve gear, carbs, camshafts and gearboxes.

TUNING STANDARD TRIUMPHS up to 1300 cc.   012.7. 50p
Richard Hudson-Evans
Essential reading for Herald, Spitfire, 1300, Standard 8 and 10 owners. Full tuning information.

TUNING STANDARD TRIUMPHS over 1300 cc.   029.1. £1.60p
David Vizard
The tuning stages for Vitesses, GT6, TRs and all 2000 units from stage 1 to full race.

TUNING VOLKSWAGENS.   Peter Noad   026.7. *£ 1.00*
An expert guide to the race and rally preparations of VWs; it covers the various types of car and their development and competition history. Includes a section on Beach Buggies.

TUNING ESCORTS AND CAPRIS.   David Vizard   009.7. *£ 1.00*
The technical editor of **'Cars and Car Conversions'** explains engine and chassis tuning procedures for both road and track.

TUNING ANGLIAS AND CORTINAS.   003.8. 80 p
This bestseller deals with engine and chassis tuning and details the early Classic and Capri, the V4 and Twin Cam power units.

TUNING TWIN CAM FORDS.   David Vizard   007.0. 80 p
The stage-by-stage modifications for these engines, from 'warm' 1600s to full-race 1800s. Fully illustrated.

TUNING BMC SPORTS CARS.   Mike Garton   004.6. 80p
The author, once a technical expert at British Leyland Special Tuning Department passes his wealth of experience on to the interested owner.

TUNING IMPS.   Willy Griffiths   052.6. 50p
The Imp is one of the most difficult cars to modify. The author lets out all the secrets on what can and cannot be done. **NEW 2nd Edition.**

TUNING VIVAS AND FIRENZAS Blydenstein & Coburn 064.X *£ 1.00*
Written by the country's leading experts, this is the first tuning book on these popular cars. It covers all aspects of tuning for both road and track.

TUNING V8 ENGINES.   David Vizard   028.3. £1.50p
This book covers the principles involved for modifying a large selection of V8 engines— design trends, supercharging, assembly, part swapping, carburation etc.

TUNING SIDE VALVE FORDS.   Bill Cooper   005.4. 80 p
This book covers the 100E engine fitted to early Ford Anglias and Prefects now finding their way into many youthful hands.

BUILDING AND RACING AN 850 MINI.   Clive Trickey. 010.1. *£ 1.00*
Another winner from Clive Trickey who describes here the story of his own racing success in a step-by-step method that can be followed by the would-be racer.

Carburetter guide

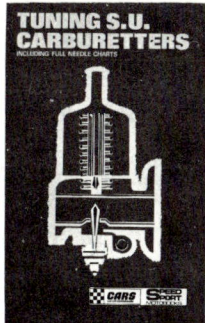

TUNING SU CARBURETTERS. 017.8. 70p
The SU carburetter is fitted to all BLMC cars and is often used for carburetter conversions to tuned cars. This book is a complete guide to their tuning, servicing and fitting, with recommended jets and full needle charts, both for the enthusiast and economy-minded motorist. Recommended by the Manufacturer.

WEBER CARBURETTERS. John Passini. 018.6. 70p
This book covers the setting-up, method of operation and servicing on one of the finest carburetters available for high performance engines. Written by an acknowledged specialist on these carbs.

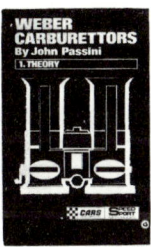

TUNING STROMBERG CARBURETTERS. 006.2. 70p
A similar volume to the SU carburetter book but for tuning the very popular Stromberg carburetter. Again recommended by the Manufacturer.

WEBER CARBURETTERS
Part 2 - Tuning and Maintenance   John Passini   060.7   70p

This is the companion volume to the very successful Weber Carbs book by the same author, which dealt with the Theory of how Webers worked and functioned only. John Passini has worked very closely with the factory to provide in this book all there is to know about Weber tuning and maintenance. Profusely illustrated and complete with needle, settings and application data tables.

and soon to follow: TUNING SOLEX CARBURETTERS

TUNING COMPANION SERIES

TUNING LUCAS IGNITION SYSTEMS   **063.1**   **£ 1.00**

This book examines each component in the Lucas ignition system
and explains how to test and check that it is functioning correctly.
Also dealt with are the special procedures and requirements of
systems on high performance engines, with setting up instructions,
trouble shooting hints and comprehensive data tables.

INTRODUCTION TO TUNING. 002.X. 50p
ENGINES AND TRANSMISSIONS. 013.5. 50p
SUSPENSIONS AND BRAKES. 027.5. 50p

Martyn Watkins has written a basic guide
to the tuning and modification of
production cars. These three volumes of the
TUNING COMPANION series are designed
to take the beginner through the theory
and then the practice stage by stage.
They should then lead him into the more
detailed work featured in the rest of the Motobook range.

AUTO ELECTRICS.   **David Westgate.**   **014.3.**   **£ 1.00**
A well illustrated and easily readable guide to the car's electrical system.
This book should be a standard work as it covers all aspects of this
complicated subject from batteries to ammeters.

CAR CUSTOMISING.   **Paul Cockburn.**   **031.3.**   **90p**
A new book on this increasingly popular form of car modification.
Paul Cockburn a brilliant young designer explains the ground work and
suggests many practical ideas.

MODIFYING PRODUCTION CYLINDER HEADS.
Clive Trickey. 008.9. 50p
Clive Trickey's famous basic guide to the modification of cylinder heads for
improved performance. A standard work which has become a best seller.

RACING ENGINE PREPARATION.   **Clive Trickey.**   **015.1.**   **£ 1.00**
Fully describes the conversion of mass-produced engines to full blown
racing units.

New edition
Castrol Book of Car Care

SBN 902-587-005
This is the new edition of the ever popular Castrol Book of Car Care in a new format and at a new price.
'Car Care' has been rewritten and considerably updated, with new drawings, diagrams and photographs. It now has a full-colour cover.
'Car Care' does exactly what its title suggests under the following chapter headings:
1. A Happy Partnership? 2. Servicing; 3. Bodywork; 4. Engine; 5. Transmission; 6. Brakes; 7. Suspension and Steering; 8. Tyres; 9. Electrics; 10. Breakdown Trouble-shooter; 11. Safety and Security; 12. Castrol at your Service.

25p

The Big Drive

THE BIG DRIVE.
Richard Hudson-Evans
and Graham Robson.
032.1. 50p
The Book of the World Cup Rally, 1970.
The first behind-the-wheel view of the toughest rally ever—the car breaking London to Mexico Race.

Castrol book

'TWO WHEELER CARE'
a sister publication to the **'Castrol Book of Car Care'** describes the various parts of the machine tells what they are designed to do and suggests the best course of action for looking after them. Used intelligently it can save a lot of time, money and frustration.
Still very popular and a constant best seller.

25p

HOW TO KEEP YOUR VW ALIVE. John Muir. **£2.50**
'A manual of step by step procedures for the complete idiot'
Softback, ring-bound, profusely illustrated

This brilliant book has been a huge success in America. It is written by an expert engineer who appreciates that complex technical procedures cannot be followed by the amateur mechanic. He explains how to look after a VW in simple language combined with a wry humour.
Basically a manual, but very different from any others. Extremely valuable even though unusual in approach. It is proving that the success in America was no flash in the pan.

SPEED SPORT
AEROBOOKS

For the Aviation Enthusiast

Hot Air Ballooning

Would you like to fly in my beautiful balloon? This catchy little tune must often have passed through people's minds as they stood and watched one of the beautiful modern Montgolfier's sailing through the air above their heads. Hot air ballooning is the oldest form of flight known to man, and is currently enjoying a tremendous renaissance. Since, however, this is such a 'new' aerial sport many potential balloonists find themselves unable to find out enough about it. This book will show the how, the why, and the wherefore of joining-in, helping, buying, and flying one of these lighter-than-air machines. Because ballooning is basically so simple, it is a reasonably easy matter to set out a pattern, which if followed, will lead the novice from stage to stage until he, or she, can become the owner-cum-pilot of one of these magnificent toys.

S.B.N. 85113-036-4   by Christine Turnbull   £1.50

Open Cockpit

In 1930, when he was eighteen years old, the author was granted a short service commission in the Royal Air Force and after completing the then one year's flying and ground subjects training he was posted to the famous 25(F) Squadron at Hawkinge. It was here, except for a few months in the Fleet Air Arm, that he spent the remainder of his five year commission flying Siskins and Furies, both single-seater fighters. After completing a Central Flying School course he joined the De Havilland School of Flying as an instructor in 1935 and of the 120 odd pupils he taught to fly at Hatfield, at least a couple were destined to become famous Battle of Britain pilots.

S.B.N. 85113-040-2   by John Nesbitt-Dufort D.S.O.   £ 1.00

Scramble

In 1941 he was awarded a DSO and later a Croix de Guerre with silver palm. Due to a certain amount of luck he escaped capture when he crashed in France in 1942 and on return to England was given a test pilot job as an operational 'rest'. He graduated on completion of the very comprehensive course at the Empire Central Flying School in 1944 and finished his second stint in the RAF at the end of 1945 commanding a mixed Spitfire and Mosquito Wing in Norway.
Now forty years since he first soloed and after 10,000 hours flying on 104 types of aircraft, he is more than qualified to discuss some of those planes, putting them into the context of their time and describing their flying peculiarities. In these books he presents a cameo of each aircraft, written in a delightful humourous manner full of personal reminiscence which will fascinate the reader. The books are neat combinations of interesting information and sheer entertainment.

S.B.N. 85113-041-0   by John Nesbitt-Dufort D.S.O.   £1.00

General Motoring

"I imagine the signs will bring quicker reaction from fellow motorists on a motorway than any amount of gesticulating or pitiful messages written on the back of an envelope."
Bradford Telegraph.

".....this could prove to be a big selling line."
Auto Accessory International.

EMERGENCY SIGNS FOR MOTORISTS
ISBN: 0-903192-08-X Size: 11¾ '' x 8¼''

A book of easily-recognizable poster-size emergency signs which can save the user time and embarrassment by showing passing motorists that something is wrong. There is a sign appertaining to almost all situations — everything, in fact, from an 'ON TOW' notice to a hazard warning sign. Perhaps of most importance is the inclusion of signs relating directly to accident prevention and the need for medical assistance. A sign such as 'DOCTOR WANTED', displayed with a large cross, attracts immediate attention and is easily understood.

"Useful when you break down"
Sunday Telegraph.

Officially approved by the Design Centre and featured on the television show **"Drive In"**, EMERGENCY SIGNS FOR MOTORISTS provides a service which is long overdue and which, as our roads grow ever more crowded, should become an essential item in the responsible driver's equipment. Remember: It will be worth more than its price when you need it!

"Quite a bright idea really"
'Drive In' Thames Television Motoring Programme

"Should be on every motorist's parcel shelf."
The Times

INTERAUTO

Workshop Series

A range of books on important but much-neglected aspects of automotive technology for the engineer and mechanic.

PETROL FUEL INJECTION SYSTEMS

ISBN: 0-903192-20-9
Size: 8½" x 11"
380 pages Illustrated

One of the first books published containing detailed information on the construction and operation of most of the major petrol fuel injection systems available today. The opening section deals with the development of the first P.I. Systems, dating as far back as 1940. This is followed by descriptive information and technical data on various systems available on the

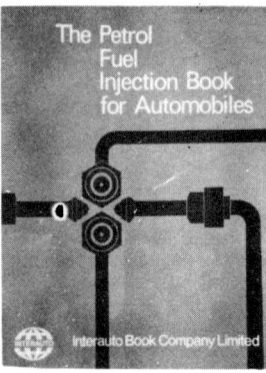

present day market.
Finally, service information on a number of vehicles to

which a P.I. System has been fitted.

With an abundance of clearly laid-out photographs, drawings and plans, and in the same large format as the other titles in the series, this book covers: AE BRICO, BOSCH(Mechanical and Electronic), KUGEL-FISHER, LUCAS & TECALEMIT in relation to the motor vehicles equipped with these systems.

ALTERNATOR SERVICE MANUAL

ISBN: 0-903192-28-4
Size: 8½" x 11"
250 pages Illustrated

This valuable publication for automotive electricians deals extensively with the testing and maintenance of Alternators and Regulators. Compiled from genuine manufacturers' service manuals.

CONTENTS: Alternator technology Bosch, Butec, CAV, Chrysler, Delco, Remy, Email, Fiat, Ford, Hitachi, Leece-Neville, Lucas, Mitsubishi. Motorola & Prestolite Application tables listing current vehicles and their standard alternators, for easy cross reference.

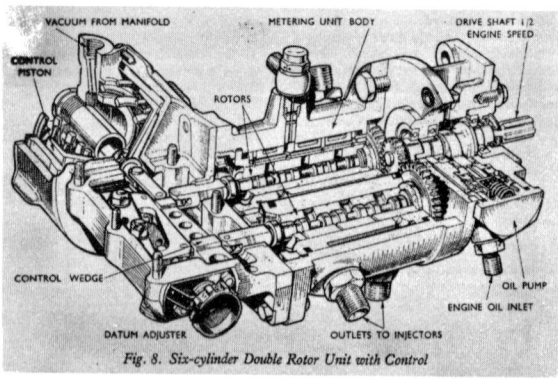

Fig. 8. Six-cylinder Double Rotor Unit with Control

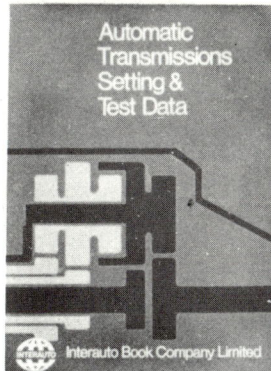

Automatic Transmissions Setting & Test Data

Interauto Book Company Limited

Crypton Triangle (Transervice) Publications

AUTOMATIC TRANS-MISSION SETTING AND TEST DATA
ISBN: 0-903192-29-2
Size: 8½" x 11"
150 pages

Presented in a compact and easy-to-read format are the setting and testing procedures for the more popular automatic transmission systems in their adapted form for use in the majority of vehicles. Additionally, the book contains such information as pressure tables, shift speeds, the location of pressure take-off points, plus comprehensive fault diagnosis charts which enable the user to carry out checks and adjustments with speed and accuracy.

The subject of automatic transmission is a complex one. This publication does not purport to be a workshop manual dealing with system overhaul and repair, but it will prove of great value to the service engineer involved in the final on-car setting and testing.

Transmission lists by vehicle make and model as cross-reference.

'Engine and Electrical Service' over 250 pages and 450 illustrations   £ 2.50

'Corrective Service' a new approach to fault finding that interprets all oscilloscope traces and meter indications   £ 2.50

'Diagnostic Wallchart' 40"x30", for quick reference, it shows all oscilloscope traces and related fault conditions   £ 2.50

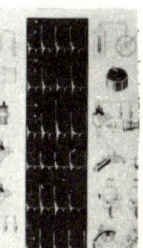

WALL CHART
40" x 30" three colour chart showing all oscilloscope traces. Ideal for checking "fault" conditions.

Interauto Books
for the professional

AUTOMOBILE FAULT DIAGNOSIS

All priced at **95p**

Interauto Automobile Engineering Reference Series

These books are directed at the car mechanic, the apprentice, the technician and the more experienced do-it-yourself motorist. In clear language, and with numerous illustrations, each title fully details a specific motor engineering subject. All books written by leading motor engineering experts and revised in accordance with 1971/2 technologies.

Part of the series was originally published in Germany by Vogel Verlag, one of that country's largest technical publishers, who have already sold more than 100,000 copies. Licensed editions of the series are also published in Spanish and Dutch.

Each book is designed to be carried around for constant and immediate reference and its handy format facilitates this. Research with major technical booksellers has shown that no reference series of this type has previously been available and that there consequently exists an unlimited demand and sales potential. A further advantage is that the series covers a truly international selection of vehicle systems.

Current titles in this series:

Automobile Fault Diagnosis
Automobile Radio Interference Suppression
Automobile Body & Paintwork Repairs
Automobile Engine Testing
Automobile Performance Testing
Automobile Diagnostic Testing

Automobile Braking Systems
Bosch Electrical Systems
Caravans - Function, Servicing, Repairs

S.U. Carburetters — Testing, Servicing, Overhauls
Zenith Carburetters — Testing, Servicing, Overhauls
Stromberg Carburetters — Testing, Servicing, Overhauls
Solex Carburetters — Testing, Servicing, Overhauls

(other carburetter books are in preparation)

All your motoring books from ONE source

ALBION SCOTT LIMITED

are the sole distributors of all
SPEEDSPORT and INTERAUTO books.

Albion Scott also distribute a range of
over 1,000 other titles for the motor
vehicle.

We call them all MOTOBOOKS

Motobooks are over 1000 Workshop Manuals,
handbooks, tuning, maintenance and repair
nooks covering practically every car on the
road today. Books on racing, veteran cars,
motor sport, biographies, connoisseur books
and many more

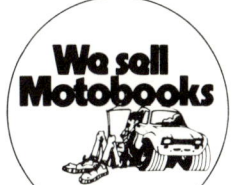

Where ever you see this sign   displayed, i.e. by good bookshops,
motor accessory shops or other retail outlest you can be sure that SpeedSport,
Interauto and the many other Motobooks distributed by us are for sale.

However, if you have difficulties in finding a suitable outlet you may order
directly from us using the form provided in this book.

Nod, Nod, Wink, Wink, Say no more...

These books are for guys like you! Full of pics of the hot ones — by popular demand we include Continental and Oriental models as well as the more usual British types (but no need for phrase books with these little beauties — know what I mean?) Everything from a simple screw to a full strip. Whether you are inexperienced or a pro, you will get more pleasure on the job after reading Motobooks! So don't delay, fill in the coupon below for our catalogue, sent in a plain, sealed envelope.

Please send me your catalogue, I enclose 10p P.O./stamps
I am over 18

Name _____

Address _____

_____ CC2

Albion Scott Ltd., 51 York Rd., Brentford, Middlesex, TW8 0QP.

HOW TO ORDER
Motobooks

Whenever you wish to purchase any of the listed books take this form to your Bookseller or Motorshop who will order the book for you. If this is not possible, mail the order form to us with your payment and we will send the required books to you by return.

Please observe the following instructions:

ALBION SCOTT LTD.·
Bercourt House
ORDERING   51 York Road
BOOKS   from   Brentford Middx
BY MAIL   TW8 OQP England

Identify required books on this form.
Mail complete form to us, with your remittance (either cheque, postal order or cash) to which you must add the postage as set out below.

Make sure that your
NAME and ADDRESS is given in the space below.

Postage and Packing:

		UK	EUROPE	OVERSEAS
Book price to	£2.00	10p	15p	20p
	£3.00	15p	20p	25p
over	£3.00	20p	30p	40p

Dispatch by surface book mail only.

Name ....

Address ....

Special Instructions ....

Get your facts straight from a Motobook

ORDER FORM

SPECIAL TITLES FROM ALBION SCOTT

Qty	Title	Price	Total		Qty	Title	Price	Total
	SPEEDSPORT					**INTERAUTO**		
	Tuning SU Carburetters	70p				Fault Diagnosis	95p	
	Tuning Weber, Part 1	70p				Interference Suppression	95p	
	Tuning Weber, Part 2	70p				Body and Paintwork	95p	
	Tuning Stromberg	70p				Performance Testing	95p	
	Tuning the Mini	£1.00				Engine Testing	95p	
	More Mini Tuning	£1.00				Braking Systems	95p	
	850 Mini	£1.00				Bosch Electrical Systems	95p	
	Four Cylinder Fords	£1.00				Diagnostic Testing	95p	
	Anglias and Cortinas	80p						
	Tuning Twin Cam Fords	80p						
	Tuning Side Valve Fords	80p				SU Carburetters	95p	
	Escorts and Capris	£1.00				Solex Carburetters	95p	
	Tuning Vivas & Firenzas	£1.00				Zenith Carburetters	95p	
	Tuning BMC Sports Cars	80p				Stromberg Carburetters	95p	
	Triumphs to 1300cc	50p				Weber Carburetters	95p	
	Tuning the VW	£1.00						
	Tuning Imps	50p						
	Tuning V8 Engines	£1.50				Caravans	95p	
	How to Start Rallying	£1.00				Alternator Manual	£2.50	
	Barry Lee Hot Rod	£1.00				Automatic Transm Data	£2.50	
	HS Motor Racing	80p				Petrol Fuel Injection	£3.80	
	How to Start Autocross	80p						
	Prod. Motorcy Racing	£1.00				Better Motoring Yearbk	95p	
	Introduction to Tuning	50p				Motorist Emerg Signs	75p	
	Engines and Transm	50p						
	Suspensions and Brakes	50p				**CRYPTON**		
	Auto Electrics	£1.00				Engine and Electrical	£2.50	
	Lucas Ignition Systems	£1.00				Transistors in Motor	95p	
	Modif. Prod. Cy Heads	50p				Corrective Service	£2.50	
	Racing Engine Prep	£1.00				Oscilloscopes in Engine	95p	
	Car Customising	£1.00				Diagnostic Wallchart	£2.50	
	High Speed Driving	£1.50				WORKSHOP MANUAL		
	Cylinder Head Modific	£2.40				quote make, model & year	£ 2.00	
	Scramble	£1.00				HANDBOOK		
	Open Cockpit	£1.00				quote make, model & year	75p	
	Hot Air Ballooning	£1.50						
	Touch Wood	£1.40				Better Motoring Yearbk	95p	
	Keep Your VW Alive	£2.50				Motorist Emerg Signs	75p	
	The Big Drive	50p						
	Castrol Bk of Car Care	25p						
	Motorcycle Care	25p						
	Qty TOTAL Price					**Qty TOTAL Price**		

NOTES. _____

_____